Planning, Implementing, and Evaluating Critical Pathways

Patricia C. Dykes, R.N., M.A., is currently working as a private health care consultant and provides services ranging from staff education to accreditation review and critical pathway design, implementation, and evaluation. In addition, she is a member of the adjunct faculty in the School of Nursing at Fairfield University in Fairfield, CT. In the past, she has held both clinical and administrative positions in medicine, surgery, and psychiatry. She holds a B.S.N. from Fairfield University and an M.A. from New York University.

Kathleen Wheeler, Ph.D., C.S., A.P.R.N., is an Associate Professor and the Director of the Graduate Program in the School of Nursing at Fairfield University School of Nursing in Fairfield, CT. She started the Psychiatric Home Care Program at United Home Care of Fairfield, CT and is the Coordinator of that program. Dr. Wheeler is a Clinical Specialist in Psychiatric Nursing and is also certified in psychotherapy and psychoanalysis and practices at The Psychotherapy Center in Fairfield, CT. She holds a B.S. from Cornell University–New York Hospital School of Nursing, and an M.A. and a Ph.D. from New York University. She is the author of numerous publications and has conducted research in outcome measurement and instrument development.

Planning, Implementing, and Evaluating Critical Pathways

A Guide for Health Care Survival into the 21st Century

Patricia C. Dykes, MA, RN
Kathleen Wheeler, PhD, CS, APRN
Editors

Springer Publishing Company

Springer Publishing Company, Inc.
536 Broadway
New York, NY 10012-3955

Cover design by Margaret Dunin
Acquisitions Editor: Ruth Chasek
Production Editor: Jeanne Libby

97 98 99 00 01/5 4 3 2 1

Planning, implementing, and evaluating critical pathways : a guide for
    health care survival into the 21st century. / Patricia C. Dykes,
    Kathleen Wheeler, editors.
        p.   cm.
    Includes bibliographical references and index.
    ISBN 0-8261-9790-6
    1. Nursing care plans.   2. Critical path analysis.   3. Medical
protocols.   I. Dykes, Patricia C.   II. Wheeler, Kathleen, Ph.D.
    [DNLM: 1. Critical Pathways.   W 84.7 P712 1997]
RT49.P56   1997
362.1'73'0684—dc21
DNLM/DLC
for Library of Congress   97-16500
                                                              CIP

Printed in the United States of America

To my parents, Arthur and Dorothy Capeci, whose wisdom, strength, and integrity set the standard for my own family.

To my husband, John, whose love, devotion, and support is the foundation of my independence.

Patricia C. Dykes

This book is dedicated to my husband, Bob Broad, whose presence and love hold my life in place, and also to my parents, Betty and Sid Wheeler, whose love gave me the strength and freedom to find my own way in life.

Kathleen Wheeler

Contents

Contributors

Barbara J. Lesperance, M.S.N., C.S., C.C.R.N., is a Clinical Specialist at Danbury Hospital in Danbury, CT. She is involved in developing critical pathways, inservicing, and outcomes analysis. She has 30 years of acute care experience, including 12 years in critical care.

Joanne Sheehan, J.D., R.N., is a Principle in the law firm of Friedman, Mellitz and Newman, P.C. in Fairfield, CT. Her concentration of practice is in personal injury, medical malpractice, and products liability litigation.

Debra A. Slye, M.N., R.N., is a Product Manager of Clinical Marketing for Multidisciplinary Critical Pathways in Kirkland, WA. She has been involved in the field of case management and information systems for over 10 years. She has published and presented extensively on the topic of automating critical pathways and clinical information systems.

Gayle H. Sullivan, J.D., R.N., is President of Quality Assurance Associates, Inc. in Fairfield, CT and a member of the Editorial Boards of *RN Magazine, Office Nurse,* and *Business & Health.* She works as a medical liability consultant and conducts educational seminars and loss control surveys for hospitals, nursing homes, ambulatory surgery centers, and homecare agencies.

Jane Woolley, Ed.D., R.N., is Director of Nursing Education, Standards, Research and Quality Assurance at the Masonic Health Care Center in Wallingford, CT. She has been involved in the design, implementation, and evaluation of critical pathways in acute care, long-term care, and rehabilitation since 1990.

Foreword

Critical pathways, the health care industry's version of project/time management tools, have passed their twelfth birthday! Despite early predictions of the demise assigned to fads, critical pathways have taken root and proven useful in countless applications. This is because critical pathways were conceptualized from the beginning to support the scientific method used by expert clinicians, i.e., assess, plan, intervene, evaluate. As this text describes, those fundamental processes can be captured on paper by a multidisciplinary author team, adapted to individual patients through copathways and algorithms, and the care that is planned can be evaluated using a combination of measurable outcomes balanced by variances encountered at the individual and cross-patient level.

While reading, keep in mind that critical pathways are only a tool, a means to an end. Although there are many ends suggested in the text, the energy behind successful implementation has to stem from health care professionals' commitment to care that is precise in its content, effect, and timing. Dykes and Wheeler understand both the commitment involved and the need for practical methods to build tools and measure their impact. They challenge us to be more patient-centered by developing pathways beyond acute care. And much to their credit, the authors give us valuable guidance about liability issues and the ultimate solution of computerization. The information and illustrations throughout PLANNING, IMPLEMENTING, AND EVALUATING CRITICAL PATHWAYS will assist the reader in creating more pathway birthdays and in positioning health care systems for the 21st century.

Karen Zander, R.N., M.S., C. S., F.A.A.N.
Center for Case Management
South Natick, Massachusetts

Preface

R ecent cost constraints in all health care settings have spurred interest in
rethinking the way care is delivered. In response to these constraints, an innovative tool, critical pathways, has emerged as a successful vehicle to deliver cost-
effective, quality care. Critical pathways define the optimal schedule of key
interventions done by all disciplines for a particular diagnosis or procedure,
designed to achieve desired patient outcomes. The genius of critical pathways lies
in their ability to cut costs without compromising on quality.

*Planning, Implementing, and Evaluating Critical Pathways: A Guide for Health
Care Survival into the 21st Century* was written to assist health care providers in
a variety of settings with assessing, planning, implementing, evaluating, and ben-
efiting from critical pathways. As the health care system enters the 21st century
and a continuum of health care evolves, seamless care across settings is possible.
However, for the sake of clarity, each chapter of this book focuses on critical pathways in a selected setting. With the exception of home care, there is very little documentation available regarding critical pathways in nonhospital settings. When
looked at logically, many of the same goals and desired outcomes are true across
settings. In the next few years, research will become available to trumpet the success of critical pathways in other health care settings. In addition, critical pathways
will emerge that address a continuum of care, rather than a single episode of care.
For example, a critical pathway written for a person with the diagnosis of pneumonia will outline care given across the appropriate continuum: primary care, acute
hospitalization, and home care follow-up. This type of pathway would contain
algorithms that would allow for professional judgment in terms of the appropriateness of a health care setting, given the patient's symptoms and desired patient
outcome.

Chapter 1, An Introduction to Critical Pathways, gives an overview of the evolution of critical pathways and how critical pathways have evolved to meet the
needs of a rapidly changing health care system. Chapter 2, Designing and Implementing Critical Pathways: An Overview, describes the process of developing critical pathways, common obstacles, and troubleshooting strategies for overcoming
problems. Chapter 3, Copathways and Algorithms, discusses how to address
comorbidities through copathways and how to integrate algorithms into pathways
to assist in clinical decision making. Chapter 4, Data Collection, Outcomes Mea-

surement, and Variance Analysis, addresses how to make use of the data that are collected through documenting on the critical pathway to improve quality of care, to document outcomes, and to use excellence as a benchmark. Chapter 5, Critical Pathways in the Acute Care Setting, focuses on developing, implementing, and evaluating critical pathways in hospitals. Although critical pathways were originally designed to be used in acute care, they have changed as the health care system has evolved. This chapter will assist the reader in learning from the mistakes of the "pioneers" and incorporating these lessons into the overall plan when setting up critical pathways. Chapters 6, 7, and 8 deal with using critical pathways in ambulatory care, home care, rehabilitation and long-term care, respectively. The authors have used the same principles that were developed in acute care for working with and developing critical pathways, but have incorporated their own expertise and that of others working in these fields to outline essential groundwork for making critical pathways successful in each setting. Liability Issues in Development, Implementation, and Documentation of Critical Pathways is addressed in chapter 9. Because of the novelty of critical pathways, there is very little literature that addresses the legal implications of critical pathways. The authors of this chapter (both practicing litigation attorneys) have outlined the advantages as well as the concerns associated with critical pathways and documentation in a court of law. Chapter 10, Critical Pathways and Computerization: Issues Driving the Need for Automation, describes some of the advantages of developing and utilizing computerized critical pathways and the possibilities that are opened up when paper-based critical pathways are automated to meet the needs of patients across a health care continuum.

It is our hope that the information contained in this book will assist groups of individuals in many health care settings with creating critical pathways in their own milieus. By using the information contained in these chapters to do the necessary preparation, it is possible to generate critical pathways that will be relevant and beneficial in any practice setting. We believe that individuals from a selected setting are in an ideal situation for creating critical pathways for that setting. These individuals are equipped with knowledge of their system, an understanding of the people that make up that system's structure, and a realization of what works and what does not in their unique setting. The most experienced and knowledgeable consultant is lacking this information and for that reason cannot achieve what individuals from inside the institution, with the proper information, can accomplish.

Patricia C. Dykes
Kathleen Wheeler

1

An Introduction to Critical Pathways

Patricia C. Dykes

THE CHANGING HEALTH CARE ENVIRONMENT

The health care delivery system of the United States is changing rapidly. In the past, health care was simple. The entire structure consisted of the physician's office and the hospital. When one became ill, one went to see one's doctor. If necessary, the physician admitted the person to a hospital. The person stayed in the hospital until the physician (and usually the patient) agreed that the illness was significantly improved. It was the same scenario, regardless of the circumstances of admission— acute or chronic illness, accidents, maternity, or substance rehabilitation. Irrespective of the length of stay, the quality of the care, and the success of the treatment, the patient's insurance company always picked up the tab, with no questions asked. Retrospective reimbursement left little incentive for providers to control cost or to monitor resource utilization. As the cost of providing health care rose to a level where it began to employ a significant portion of the U.S. gross national product, it became evident that if measures were not taken to slow the growth of health care cost, Medicare and Medicaid would run out of funding early in the 21st century. This revelation started a chain of events that have left health care in a constant state of flux.

Over the last 30 years, the picture of health care has changed dramatically. In 1960, the cost of health care accounted for 5.3% of our GNP (U.S. Department of Health and Human Services, 1991). By the mid-1970s, this number increased to 8.3% (U.S. Department of Health and Human Services). In 1991, the cost of health care jumped to 13.2 % of the GNP (U.S. Bureau of the Census, 1994). As health care costs continue to consume a greater portion of the GNP, the number of people with health care insurance is decreasing. In the period between 1980 and 1989, the number of uninsured persons under age 65 increased from 12.5 % to 15.7% (U.S. Department of Health and Human Services).

The health care system in the United States no longer consists of the physician's office and the hospital. That basic, two-tiered system has evolved into a health care continuum consisting of ambulatory care clinics, surgi-centers, physicians' offices, outpatient rehabilitation, hospitals, subacute care centers, long-term care facilities, and home care services. In 1980, 16% of hospital surgeries were done on an outpatient basis (U.S. Bureau of the Census, 1994). By 1992, outpatient surgery accounted for 54% of hospital surgeries (U.S. Bureau of the Census). Today, most health care services are delivered in alternative settings, with a preponderance provided on an outpatient basis.

WHY CRITICAL PATHWAYS?

Critical pathways make the most sense when they are considered as part of the natural evolution of health care. In the past, it was sufficient for hospitals to provide what the hospital considered to be "good care." No one ever asked about outcomes. As technology improved, it became possible to provide treatments that a short time ago were inconceivable. Initially, we gazed at these treatments with amazement and awe. Marvel has since been replaced with trepidation as we consider the price of technology in terms of both finances and human suffering. The cost of modern health care has reached its saturation point. We now must change the way we have traditionally delivered care, or we must begin to ration health care.

The state of health care is at a turning point. Until now, the fastest growing portion of the GNP has been consumed by a service whose quality was at the mercy of the caregiver. Medical and nursing research have been done for centuries to better understand disease and to refine specific treatments. What has not been looked at closely is the way in which care is coordinated and then delivered. Traditionally, when a patient was admitted into the health care system, the quality of care was dependent on the coordination of care delivered by many providers. In addition to scientific knowledge, the basis of much of the care was tradition and the experience of individual providers. No one was ever asked to document the effectiveness of the care that one delivered. Each discipline had its own care plan or treatment plan, but was rarely held to achievement of the goals and outcomes outlined in those plans. As the cost of health care has continued to escalate and people are responsible for a greater portion of their health care costs, more attention is being paid to the quality of the health care received. An increase in the cost of malpractice cases and a cumulation of accrediting agencies and consumer watchdog agencies looking directly at health care have spurred on a new sense of accountability.

Changes in Health Care Reimbursement and the Emergence of Critical Pathways

The unraveling of the traditional system largely began in the 1980s when Medicare and Medicaid changed from retrospective to prospective systems of reimbursement. Insurance companies followed suit, and for the first time in history, the profits associated with delivering health care dropped, while the costs continued to rise. Many physicians and hospitals were caught off guard and considered this a short-term interruption that would end in disaster for the patient. Many reasoned that these changes were temporary; the government and insurance companies would see the havoc of their policies and would revert to a "reasonable" system of reimbursement. Others took this foreshadowing seriously and began to look at the way they were delivering health care. Like other businesses, these agencies looked for ways to cut out waste and to reduce costs without compromising quality.

New England Medical Center was one such place. They had the foresight to take these reimbursement changes seriously and the impetus to act on them. It was from the New England Medical Center that the first critical pathways emerged. These were formulated as a part of a larger case management plan. The general assumption was that patients were admitted to hospitals largely for nursing care and technology. By delivering only the necessary technology (technology is expensive) and concentrating only on nursing actions associated with achieving predetermined outcomes, the length of stay could be shortened and the associated costs, therefore, would be diminished. It was believed that by creating a master plan for each diagnosis and then an abbreviated care map (the critical pathway), nursing care could then be intensified and patients could be discharged in a shorter period of time (Zander, 1988a). To delineate this, multidisciplinary teams gathered to formulate outcomes for each diagnosis. Under each diagnosis, key interventions aimed at realizing the stated outcomes were categorized by general classifications or care elements. These included consultations, assessments, treatments, nutrition, medications, activity, safety, teaching, and discharge planning. As a result of this health care delivery experiment, New England Medical Center was put on the map as a center of progression. The medical center reported numerous positive outcomes related to their case management plan and the use of critical pathways (Zander, 1988b).

Many of these claims have since been substantiated. In the last few years, the literature has mushroomed with articles pertaining to managed care, critical pathways, and the benefits of instituting both during the reorganization of health care. Benefits include new ways to limit costs and waste, while maintaining and improving quality. Accomplishment of these long-term goals requires implementation of a tool that builds on the most preeminent concepts of the day: case management (CM), continuous quality improvement (CQI), and total quality management (TQM). Critical pathways are this tool.

Critical Pathways, Defining Cost, and Documenting Outcomes

Unlike the period before the 1980s, all insurance companies routinely question care delivered. Health care providers seldom have the flexibility to report their costs and expect full or even partial reimbursement. HMOs and Managed Care Networks have emerged and account for greater amounts of the health care reimbursement pool. These networks take bids from many providers and send their customers to the provider who has the lowest cost and who can also document the highest quality of care. Rising costs, managed care networks, and public interest in health care have led to competition among health care providers to strive for quality care, while controlling cost.

Although critical pathways were originally designed to keep down costs by limiting technology and decreasing the length of stay, they have evolved into a tool that achieves these outcomes as well as others. When set up properly, critical pathways represent a multidisciplinary plan for the patient. With all disciplines following the same plan, the patient's care is truly comprehensive. Critical pathways cut out unnecessary tests and procedures, standardize care that is given, and provide documentation of hard core data to use for quality improvement. With the data collected, health care agencies are now able to study the effects and significance of the care that they deliver. For the first time, health care agencies can document the merit of their multidisciplinary "product."

In addition, critical pathways make it possible to cost out services for treating specific populations. This is very positive for the nursing profession, which has been looking for ways to define its practice in terms of cost and outcomes for decades. The critical pathway outlines all of the tests, procedures, treatments, and teaching services that patients utilize during their length of stay. This is true for hours of stay in an ambulatory care clinic or intensive care unit, days of stay in a hospital or subacute care center, or visits to patients receiving home care. By adding up the cost of the services outlined in the critical pathway, an agency can come up with a rough cost for specific diagnoses in a given setting, or over a period of time in a continuum of settings. Because any additional services the patient receives that are not on the critical pathway are recorded as variances, these too are tracked and can be figured into the total cost of the service.

Critical Pathways as Continuous Quality Improvement Tools

The critical pathway can also be used as a continuous quality improvement tool. Documentation on the critical pathway tracks patient progress toward stated goals and outcomes. With computerized charting, it is possible to tabulate QI data while documenting. Because all of the patient outcomes are included on the critical pathway, no additional paperwork is needed. Also, variances from patient goals and outcomes are recognized in a timely manner and acted upon, when necessary. Variance patterns can then be analyzed, and the critical pathway and practice patterns can be changed where appropriate. Proper tracking requires close communication among all disciplines involved in the patient's care. This opens up lines of communication and creates an atmosphere of collegiality around the patient. Managers keep track of long-term trends such as length of stay, discharge destination, complication rates, functional status, recidivism, and the total cost of care. This information can then be used to bench mark excellence, substantiate care with insurance companies and accreditation agencies, compete for business, and improve on care delivered.

Critical pathways make it possible for health care agencies to continually undertake research and improve their product. When utilized properly, critical pathways are never finalized, as the research generated changes practice and thus the critical pathways. In essence, critical pathways provide a means for providers to promote and document their excellence so that they can remain competitive.

Critical Pathways as Teaching Tools

Critical pathways also serve as teaching tools for staff and patients. Critical pathways outline the care given for specific diagnoses and procedures. Therefore, part-time or per diem staff can care for an unfamiliar patient and give care similar to that delivered by a primary caregiver. New staff members and students can also make use of the critical pathway as an orientation tool. Critical pathways may be simplified and then used as patient teaching tools.

The patient critical pathway outlines the care that will be delivered and the expectations of the patient as a partner in that care. When recognized as a partner, it becomes more difficult for a patient to "slip through the cracks." The patient critical pathway arms patients with the information that is needed to accept responsibility for their care, their recovery, and their discharge. The patient critical pathway has ramifications for the staff as well as the patient. It is no longer acceptable to keep the patient in the dark regarding her or his care. Often, if care is not delivered as outlined on the patient's critical pathway, the patient wants to know why it is not. The staff must be aware of variances and be prepared to explain them to the patient.

CONCLUSIONS

As critical pathways evolve, they have come to play a major role in health care. Critical pathways began as a single case management tool, but have expanded to become the essence of case management as well. When used to their potential, critical pathways function as case management, documentation, teaching, quality assurance, and a host of other tools. As external forces that continue to demand quality strike a balance with internal forces calling out for cost constraint, the role of the critical pathway in the future of health care is guaranteed distinction.

REFERENCES

U.S. Bureau of the Census (1994). *Statistical Abstract of the United States* (114th ed.). Washington, DC.

U.S. Department of Health and Human Services (1991). *Health, United States and Prevention Profile.* Hyattsville, MD.

Zander, K. (1988a). Nursing case management: Resolving the DRG paradox. *Nursing Clinics of North America, 23*(3), 503–519.

Zander, K. (1988b). Nursing case management: Strategic management of cost and quality outcomes. *Journal of Nursing Administration, 18*(5), 23–30.

BIBLIOGRAPHY

Allred, C., Arford, P., Michel, Y., Veitch, J., Dring, R., & Carter, V. (1995). Case management: The relationship between structure and environment. *Nursing Economics, 13*(1), 32–41.

Capuano, T. (1995). Clinical pathways: Practical approaches, positive outcomes. *Nursing Management, 26*(1), 34–37.

Coffey, R., Richard, J., Remmert, C., LeRoy, S., Schoville, R., & Baldwin, P. (1992). An introduction to critical paths. *Quality Management in Health Care, 1*(1), 45–54.

Cohen, E., & Cesta, T. (1993). *Nursing case management: From concept to evaluation.* St. Louis: Mosby.

Crummer, M., & Carter, V. (1993). Critical pathways—the pivotal tool. *Journal of Cardiovascular Nursing, 7*(4), 30–37.

DeWoody, S., & Price, J. (1994). A systems approach to multidimensional critical paths. *Nursing Management, 25*(11), 47–51.

Dijerome, L. (1992). The nursing case management computerized system: Meeting the challenge of health care delivery through technology. *Computers in Nursing, 10*(6), 250–257.

Hammer, M., & Champy, J. (1993). *Reengineering the corporation: A manifesto for business revolution.* New York: HarperCollins.

Hofmann, P. (1993). Critical path method: An important tool for coordinating clinical care. *Journal of Quality Improvement, 19*(7), 235–244.

Ley, J. (1995). Putting critical pathways on the map. *Critical Care Nurse, 6,* 106–113.

Lumsdon, K., & Hagland, M. (1993). Mapping care. *Hospitals and Health Networks, 67,* 34–40.

Martich, D. (1992). The role of the nurse educator in the development of critical pathways. *Journal of Nursing Staff Development, 9*(5), 227–229.

McGinty, H., Andreoni, V., & Quigley, M. (1993). Building a managed care approach. *Nursing Management, 24*(8), 34–35.

Mosher, C., Cronk, P., Kidd, A., McCormick, P., Stockton, S., & Sulla, C. (1992). Upgrading practice with critical pathways. *American Journal of Nursing, 92*(1), 41–44.

Moss, M. (1994). Nursing tools: A global perspective. *Nursing Management, 25*(6), 64a–64b.

Olivas, G., Del Togno-Armanasco, V., Erickson, J., & Harter, S. (1989). Case management: a bottom-line care delivery model: The concept. *Journal of Nursing Administration, 19*(11, Pt. 1), 17–20.

Olivas, G., Del Togno-Armanasco, V., Erickson, J., & Harter, S. (1989). Case management: a bottom-line care delivery model: Adaptation of the model. *Journal of Nursing Administration, 19*(12, Pt. 2), 12–17.

Rasmussen, N., & Gengler, T. (1994). Clinical pathways of care: The route to better communication. *Nursing, 2,* 47–49.

Woodyard, L., & Sheetz, J. (1993). Critical pathway patient outcomes: The missing standard. *Journal of Nursing Care Quality, 8*(1), 51–57.

Zander, K. (1991). Care maps: The core of cost/quality care. *New Definition, 3,* 1–3.

Zander, K. (1992). Focusing on patient outcome: Case management in the 90's. *Dimension of Critical Care Nursing, 11*(3), 127–129

Zander, K., & McGill, R. (1994). Critical and anticipated recovery paths: Only the beginning. *Nursing Management, 25*(8), 34–40.

2

Designing and Implementing Critical Pathways: An Overview

Patricia C. Dykes

DEFINING CRITICAL PATHWAYS

Critical pathways define the optimal sequence or timing of the key interventions done by all disciplines involved in patient care for a particular diagnosis or procedure (Coffey et al., 1992). Critical pathways are also called care maps. Some institutions are reluctant to use the word "critical" because of the implications it may have for the patients when they learn they are on a "critical" pathway. Others are bothered by the word "pathway" because it may imply that there is one *best* way to deliver care and all other ways would be considered substandard. Although much of the documented success with critical pathways has occurred in the acute care area, critical pathways can be utilized in any setting. Many health care organizations outside of acute care are currently developing critical pathways to meet the needs of clients in other settings.

The general principles of design and implementation hold true across settings, that is, critical pathways are designed by first identifying desirable outcomes and then mapping out interventions intended to achieve those outcomes. What changes across settings is the period of care for which the critical path is developed (Zander, 1988). In acute care, subacute care, and rehabilitation, critical pathways are initiated at the time of preadmission screening or on admission and are terminated at the time of discharge. The interventions are mapped out in days of care. In ambulatory care, critical care, and emergency care, critical pathways define hours or even minutes of care. Home care providers can utilize critical pathways to delineate care delivered during each home visit. Primary care providers can create critical pathways to outline episodes of care for persons suffering from acute or chronic illness. As more health care provider networks are formed, critical pathways will be written to address specific outcomes for each diagnosis across settings. In this way, patient care and treatment will flow from one setting to the next.

Critical Pathways and Multidisciplinary Design

Developing multidisciplinary critical pathways is a major step toward coordinating and organizing care. For this reason, critical pathways should be developed by the professionals that will use them. By meeting to discuss general practice, present current literature, and define expected patient outcomes, professionals are taking a giant step toward making a commitment to consistent, high-quality care. Once this commitment is made, the quality of the patient's care will no longer be dependent on the particular staff member assigned to a patient on a given day.

With economic and consumer pressures on health care directed toward outcomes management, continuous quality improvement, patient satisfaction, and total quality management, it is easy for institutions to become impulsive and to want to "jump on the critical pathway bandwagon." Critical pathways are a change in philosophy for many institutions and should not be approached lightly. There must be significant "buy-in" from all levels in an institution for critical pathways to be successful. Like any change, there is bound to be dissension within the staff (nurses, physicians, physical therapists, dietitians, and phlebotomists). Without a firm commitment from the leadership, critical pathways will not achieve what they are designed to do, to reduce costs and length of stay while improving on quality and patient satisfaction. The most important step any institution can take toward implementing critical pathways is to take the time necessary to get buy-in within the institution. The leadership team needs to agree that a philosophical change of this magnitude and in this direction is right for the institution. If there is consensus within the leadership, the next step is to get buy-in from the staff.

Strategies for Achieving Buy-In

To get the staff committed to critical pathways is possibly the greatest challenge in any setting. All change is predictable in that one can accurately predict that change will be initially difficult and unpopular. Change associated with critical pathways will be no exception. There will be a great deal of frustration in the beginning stages of development and implementation that will need to be worked out over time. This must be done before the critical pathways can be utilized effectively and before the staff will realize the benefits. Old ways of practice need to be let go, and a grieving process must ensue. For this reason, an important step toward working through this period is for the staff to participate in the development of the critical pathways that they will be required to use.

Critical pathways are available in the literature for specific diagnoses or procedures, and some publishing companies sell critical pathways for a fee. Although it may seem like a simple solution to purchase a set of critical pathways and to just "put them into place," this does not grant staff a vehicle in which to learn about critical pathways or the time needed to work through the problems they may have with them. The staff will have many valid concerns about critical pathways before they are implemented and deserve a chance to express these concerns and to be given a forum for resolving them. Allowing staff input in development and implementation grants this forum.

Defining Practice

In addition to providing an arena for learning, expressing concerns, and working out problems, involving the staff in the creative process forces them to define their practice. Within all professional settings, there are wide variations in practice patterns. Some staff routinely read the literature and incorporate significant findings into their practice. Others rely on tradition and seldom question the status quo. Enormous growth occurs for all who participate in critical pathway development, for staff are forced to discuss as a group how care is routinely delivered. This provides an opportunity for debate and open discussion.

Before writing a critical pathway, the staff decide how to map out all of the significant consults, procedures, teaching, tests, and treatments that a patient will receive over a given length of stay. It is helpful at this point for the staff to refer to relevant critical pathways in the literature or from other institutions to use as guides when formulating their own critical pathway. The staff must agree first on which interventions are truly necessary, given desired patient outcomes. Then decisions are made regarding the occurrence of each intervention and who will provide the intervention. The group's aim is to focus on the majority of patients with a given diagnosis and to avoid being distracted by the exceptions. This can be a very threatening experience for some, because the bases of their practice may be called into question.

When the critical pathway is complete, it may not reflect the day-to-day practice of each person in the group, but a culmination of everyone's input. When the time comes for implementation, the staff will understand why specific interventions take place where they do on the critical pathway. Participation in the development provides an in-depth understanding of critical pathways and what effects they might have on patient outcomes. When the staff encounter problems with the critical pathway in practice, they will be familiar with the vehicle for working out these problems. When the staff do not have input into critical pathway development, they are more apt to question the critical pathway because it may not represent familiar practice. Standardized critical pathways give credence to the term "cookbook medicine" and augment staff feelings that the institution is attempting to dictate practice. When the staff have problems with critical pathways they did not develop, the entire pathway is more likely to be disregarded, rather than worked on in areas that need correction.

Computerized Critical Pathways

Once the staff have written critical pathways that are reflective of their practice, an institution may decide to purchase a computerized critical pathway program to simplify data collection and variance analysis. Any good program allows for individualization of critical pathways to reflect institutional practice. In this way, the staff have the convenience of computerized documentation in a format that reflects their own practice. Whether an institution purchases a computerized critical pathway program or develops its own, ultimately some type of computerization is necessary to get the maximum benefit out of the critical pathways. Critical pathways organize data and allow for significant data collection. As critical pathways evolve, they become more complex and the amount of data collected becomes limited by the person(s) collecting the data. By automating the critical pathway, endless analysis of data can be achieved.

Critical Pathway Opposition and Agency Response

Critical pathways can be used in any setting for a given diagnosis or procedure. A basic assumption behind critical pathways is the "eighty/twenty rule," that is, 80% of patients follow a predictable path 100% of the time. An additional 20% of patients stray from that pathway, and a portion of that 20% deviate far from the original pathway. Others deviate only slightly and will return to the pathway after the initial deviation. Still others start out on the critical pathway, experience a few complications (variances), and then after a period of time, return to the pathway.

It is the population that falls into this 20% category that adds fuel to the fire for those opposed to the use of critical pathways. One argument against the use of critical pathways is that they are "cookbook medicine." Opponents indignantly warn of the dangers of following critical pathways and not using clinical judgment to do what is "best" for the patient. Often when pressed, these same practitioners will agree that most patients within diagnostic categories do follow a similar pathway, but continue to express concern for the few who do not. It is important to educate all practitioners from the beginning so that they will understand that the critical pathway is not meant to be the answer for all patients at all times. The critical pathways will improve consistency of care and provide a sound guideline for the majority of patients, but are never meant to be followed blindly. Practitioners should be encouraged to question routinely aspects of the critical pathway that may not meet the needs of selected patients. Critical pathways can always be modified to meet the needs of individual patients. The final say on what is done for the patient is always left up to the team of professionals caring for the patient.

A deviation from the critical pathway is recorded as a variance. A variance is not inherently bad or unsatisfactory. It simply means that the pathway was not followed absolutely. Variance data are recorded and analyzed over a period of time. If these data show routine variances on specific areas of the critical pathway, these variances are studied in depth. Careful review might show that the critical pathway was inaccurate, and the pathway then would be changed to reflect actual practice. The review might potentially highlight practice patterns that are not consistent with what is current and acceptable in general practice and the literature. In this case, practice patterns would need to be addressed with the appropriate practitioners. In both cases, the patient is protected, and in the long run, he or she benefits from the use of the critical pathway.

Another common objection to critical pathways is that they are the institution's way of "dictating practice" in order to save money. The response to this reaction should be honest and nondefensive. Administrators need to do the proper groundwork before introducing critical pathways or any major change. Part of that groundwork is understanding the true motivation for developing and implementing critical pathways. If the major incentive is financial, this needs to be discussed openly and unequivocally. Frequently, employees look at financial incentive for change as intrinsically malevolent. This is often the case because the administration may make a change for a financial reason, but labels the motivation as something that sounds less self-serving, such as to improve patient care or to boost patient or employee satisfaction. When motivation is mislabeled, employees can often see the real incentive and subsequently feel manipulated. At this point, a common reaction is to look for ways to sabotage the planned change.

Laying the proper foundation before implementing critical pathways can prevent incapacitation. Administrators need to educate staff routinely about the financial realities of staying in business in the late 1990s and into the next century. Part of this reality includes remaining fiscally sound so that specific services can continue to be provided. If financial incentives are the primary motivation for instituting critical pathways, other cost-saving measures should be examined in addition to implementing critical pathways. Although cost savings often accompany the institution of critical pathways, there may be an initial increase in costs to cover employee time to generate, to implement, and to learn to utilize the critical pathways. In institutions already working with minimal staff and resources, there may never be a decrease in cost, but a focus instead on improving the quality of the services provided. As health care continues to become more competitive, institutions will need to provide high-quality care at a reasonable price to remain in business.

DEVELOPING THE CRITICAL PATHWAY

Getting Started

Once the administrative decision has been made to implement critical pathways, the problem becomes choosing a diagnosis or procedure to address first. Generally, diagnoses should be considered if they are high cost, high volume, high risk, or a combination of these factors. In addition, it is helpful to use data from patient satisfaction surveys and financial reports when choosing a diagnosis. It is best to choose a simple diagnosis and to use it as a pilot study when designing and implementing the first critical pathway. This will provide a learning tool and minimize frustration in the initial stages of pathway development. The first critical pathway should be simple enough to offer some guarantee of success, to reward staff for their efforts, and to inspire work on subsequent critical pathways. Generally, the surgical diagnoses tend to be more clean-cut and manageable than the medical diagnoses. It is also helpful to choose a diagnosis that is well documented in the literature and has associated, research-based practice standards that can be utilized when mapping out the care elements and interventions. This cuts down on dissension between task force members that may occur when discussing acceptable practice. It also makes key interventions under each diagnosis more definitive.

Once a diagnosis or procedure is chosen, a multidisciplinary team is formulated to write the critical pathway. The team should include members of each discipline that is relevant to the chosen diagnosis. For example, if the diagnosis selected is DRG (Diagnosis Related Group) 210 (fractured hip), it would be beneficial to have physicians (from the emergency room, orthopedics, and anesthesiology); nurses (from the orthopedic unit, utilization review, and the emergency room), and representatives from physical therapy, occupational therapy, social work, and risk management. A patient admitted with the diagnosis of CVA (Cerebrovascular Accident; DRG 14) would have a multidisciplinary team that included a neurologist, a nurse from the neuro unit and rehabilitation, and representatives from dietary, speech therapy, physical therapy, occupational therapy, risk management, social work, and utilization review. Including all of the disciplines involved in the care of patients with the diagnosis under consideration is an important step toward making all staff members feel involved in the change process and ultimately accept critical pathways.

Choosing the Team

When choosing staff members for multidisciplinary teams, it is important to appoint staff that are team players so that they are able to work as a group. It is tempting to choose staff that are generally accepting of change and are in favor of implementing critical pathways. Initially, this makes the multidisciplinary meetings more pleasant, but has a long-term potential for being less productive. If all members of the multidisciplinary team are prochange, there is a danger of creating a "group think" situation, whereby people blindly agree to implementing something that may not be practical. It is healthy to have a variety of views voiced in the multidisciplinary meetings. Chances are that the views held by committee members will be mirrored by other staff members outside the task force. By addressing concerns openly, problem areas can be resolved before the critical pathway is implemented. If the first critical pathway is implemented blindly and major problems are identified after implementation, this gives staff a distorted impression of what critical pathways are and how they are meant to function. This also makes staff reluctant to use the pathway and undermines future critical pathways.

The Group Leader

It is also helpful to have a strong group leader who is respected by the staff. Ideally it should be someone who has influence on physicians as well as the other disciplines. Many institutions use nurses as group leaders, but a joint nurse-physician leadership team is ideal, especially when addressing compliance and changes in practice within disciplines. It is the job of the group leader to create an atmosphere where people can react honestly and express ideas and concerns openly. At times, the group leader may need to redirect negative energy so that it works for the group's goals. For example, if one team member is unable to resolve concerns related to critical pathways being "cookbook medicine," perhaps this team member could be assigned to work on the actual design of the critical pathway. In this way, he or she would be able to build safeguards into the critical pathway to address these concerns. Directing the unresolved tension into productive channels helps the staff member work through personal problems with the critical pathway, and also helps to accomplish the stated goal of the group: to develop the critical pathway.

The Work of the Critical Pathway Committee

The group leader sets the tone in the first committee meeting for future meetings. For this reason, it is important for the leader to have a prepublished, set agenda for each meeting and to make each meeting a work meeting. From the beginning, group members are given the message that their time and input are respected and valuable. A strong group leader is able to help the group focus, without inhibiting creativity. The leader sets the ground rules at the first meeting, and among these rules is the stipulation that there are no "sacred cows" when looking at practice issues. If the group is to make an impact, they must be free to examine all areas of practice.

Given the time constraints of the committee members, it is not always possible to involve all members in initial chart reviews to research standard practice. If committee members are available and willing, chart reviews provide a good education into background data. The random chart review is done to determine current practice patterns and patient outcomes. This information is invaluable in terms of giving the group a focus and a starting point for creating the critical pathway. If it is not possible to involve committee members in this process, it may be possible to utilize other related staff members to do portions of the chart review. This involves a greater number of personnel in the process of critical pathway development and ultimately has the potential for greater buy-in.

It is helpful to have the data from a random chart review, a review of the literature, and any documented practice standards available at the first committee meeting. Presentation of this data in an easy-to-follow, yet professional, format (such as a slide show) provides a pivotal point for group discussion. With this preliminary information available, the group leader is able to document what the current level of practice is, what the literature and practice standards recommend, and then he or she allows the group to decide what areas of practice are of concern and how to best develop a pathway.

The first step in critical pathway development is for the group to agree on patient outcomes. This is crucial because each discipline must be clear as to what it is trying to achieve, before it maps out how it will achieve it. General outcomes that are standard across diagnoses are: increasing patient satisfaction, minimizing complications and recidivism, decreasing costs, and decreasing length of stay. Other outcomes are diagnoses or procedure specific. Information from the literature review is helpful in setting these standards.

In addition, there may be specific areas of practice that need to be addressed by the committee in the process of pathway development. If this is the case, subcommittees may need to be formed to address additional concerns. For example, if the stated objective of the group is to write a critical pathway for DRG 210 (fractured hip) and the preliminary data show a length of stay (LOS) of 14 days (DRG allotted LOS is 7 days), the group will probably decide that length of stay needs to be looked at more closely. A chart review might reveal that the emergency room (ER) LOS is 24 hours, the mean time from admission to surgery is 3 days, and the mean LOS from the time the patient is medically cleared for discharge and actually discharged is 5 days. In this case, the group has a lot of individual issues that require closer study. One subcommittee may be formed to look at ways to reduce LOS in the ER. Another group may address issues related to medical clearance. Still another subcommittee may be formed to study discharge barriers.

Group members are encouraged to go beyond simply improving the existing processes, but to look at all interventions, treatments, and processes in a fresh way. Interventions should be rethought and conventional wisdom left behind. Tradition may need to be discarded if its relevance has been lost. There are no longer resources or time to spend on interventions that do not have any basis in research and improving outcomes. There needs to be one person assigned as the chairperson for each subcommittee so that he or she can take responsibility for presenting the findings of the group at the next meeting. An agenda and goals for each subsequent meeting should be set by the end of each meeting so that each group member is aware of individual and group responsibilities. By requiring a formal presentation from the subcommittees at meetings, a time limit is set for any work that needs to be done. All of these actions involve group members and encourage them to take responsibility for development and promote long-term investment in the critical pathway.

THE CRITICAL PATHWAY AND ITS COMPONENTS

Once the multidisciplinary group has formulated the patient outcomes, it needs to settle into the task of writing the critical pathway. A review of current literature will usually provide reference for critical pathways for most diagnoses or procedures under consideration. If a critical pathway is not available in the literature, another institution may have one that it is willing to share. Otherwise, there may be one in the literature that is close enough in principle to use as a guide. Critical pathways are usually written as single-page charts, but take on a host of formats. The group may need to use a few for reference, one or two for content, and others for referencing format.

Design Considerations

When designing the critical pathway, it is important to keep in mind its function in the setting where it will be used and the outcomes it is designed to achieve. The format also depends on whether the critical pathway will be used to outline care as it is given or whether it has other documentation-related functions. Regardless of the function, the critical pathway must be legible. That means that it cannot be printed in a font that is so small it can only be seen with a magnifying glass. If simple identification and classification of variances will be done on the critical pathway, this is accomplished without taking up a great deal of space. If notes will be written on the critical pathway to explain variances, this may take up a significant amount of space and may prove to be impractical. Within the next few years, most institutions will be computerized, and this also needs to be considered when designing a format for critical pathways. Choosing a simple format that can be transferred to an automated documentation program saves time later.

In addition, any documentation that can be done on the critical pathway and can replace other forms of documentation saves staff time and decreases opposition to the critical pathway. When designed with this in mind, critical pathways have the potential for decreasing the amount of documentation, because they are a form of outcome-based documentation. When all of the essential elements of care are included on the critical pathway, only variances need to be documented.

Setting Goals

Goals are set using data from chart reviews, bench marking studies, and desired patient outcomes. These factors are considered along with institutional realities, such as mean length of stay, current practice patterns, and patient population. For example, the projected length of stay (LOS) is usually based on the reimbursable DRG LOS, but it can be more or less aggressive, depending on the institution's goals. Thus if an institution currently has a LOS that is well over the DRG recommended LOS, it might be realistic to initially set the LOS slightly above the DRG recommended LOS, but lower than the institution's current LOS. Once this intermediate goal is realized, the LOS may be lowered. If the institution's LOS for a given diagnosis is at, or lower than, the DRG recommended LOS, it might be beneficial to keep the LOS unchanged and to look at other factors within the diagnosis that might be reduced. Aspects of care vary from institution to institution or between diagnoses, but generally include the following: assessments, consultations, activity, safety, teaching, tests and procedures, nutrition, medication, treatments, and discharge planning (Zander, 1988).

The First Draft

The multidisciplinary meeting may be used as a forum to create the first draft of the critical pathway. By using a large erasable board and some markers, a blank critical pathway is drawn on the board. Each task force member is provided with a sample pathway from the literature and a blank pathway. Together, task force members complete the critical pathway. Each discipline takes a turn filling in the areas relevant to its practice. The large mock-up makes it possible for discussion and revisions. Many aspects of care will not be controversial, and are easily arranged on the critical pathway. Others will be very controversial and may generate heated discussion. Because there are many variations of "standard practice," it may be difficult to pin down the exact day or time a treatment is done over a given length of stay.

The group leader's role is crucial during this process and keeps the group focused on the task at hand. The group leader may need to remind members that critical pathways do not dictate practice and that the group needs to avoid getting sidetracked on the exceptions and focus on the majority of patients. There are times when one practitioner is uncomfortable with changing personal practice so that it conforms to the critical pathway. When this occurs, it is a good opportunity to remind the group that all practitioners will be expected to utilize clinical judgment for every patient and to do what they believe is best for each patient, regardless of what the critical pathway specifies.

The critical pathway provides a tool to study practice patterns. Over time, the outcome data furnish information about the legitimacy of practice patterns, and this information is then used to improve practice. For example, if the group is writing a critical pathway for DRG 148 (small and large bowel surgery), the physicians may not be able to agree on the appropriate day or time to remove the nasogastric tube. Suppose most of the physicians agree on removing the tube on postoperative day 1, but one or two are not comfortable with this practice. The first step would be to go back to the literature and look for documented practice standards and research. If practice standards exist, the critical pathway should be written to support these. If nothing definitive is found in the literature, it is not practical to write a separate critical pathway for each physician. The practice of the majority should be outlined on the critical pathway. Any variation in this practice will be recorded as a variance, and then long-term trends can be studied over time. Based on long-term trends, the critical pathway will be changed, if necessary.

Once the first draft of the critical pathway is complete for each discipline, it can be transferred to a hard copy and distributed to each committee member. Each discipline is then responsible for bringing the critical pathway back to staff meetings for its discipline, sharing the information and getting feedback. In addition, one or two members of each discipline should be appointed to cross-reference the critical pathway with practice standards and research to be sure that the critical pathway conforms. A target date should be set for the next meeting so that all committee members are aware of the deadlines for completing their assignments.

The feedback presented at the next meeting is used to refine the critical pathway. Once this is done, the refined version of the critical pathway should be sent out to committee members and all staff that will be utilizing the critical pathway. A memo should be attached that explains the function of the critical pathway and encourages the staff to study the pathway and to send any concerns, comments, or changes to the group leader in writing by a specified date. The memo should note that if feedback is not received by the specified date, it will be assumed that the critical pathway is in an acceptable form. The multidisciplinary committee uses this feedback to finalize the critical pathway.

Critical Pathway Accompaniments

The committee members then decide what forms, teaching tools, quality assurance monitors, algorithms or copathways, and satisfaction surveys will be implemented with the critical pathway, and how these will be developed. For example, the committee members might decide to implement a patient information booklet and a patient critical pathway to fully educate and involve patients in their care. The multidisciplinary group must look at what is currently in place for a given diagnosis or procedure. It is important to have some patient education material available to go along with each critical pathway. Educating the patient allows the patient to take responsibility for his or her care. When care is not delivered in the manner in which the patient is taught it should be delivered, he or she often asks for an explanation. This provides one more safeguard against errors and oversights.

Critical Pathways and Documentation

In addition to deciding what tools must accompany the critical pathway, the group must also determine which pieces of documentation can be replaced by the critical pathway. It is a good idea, when adding a documentation tool such as a critical pathway, also to take at least one away. In this way, the staff does not feel burdened by additional paperwork. The more user-friendly the staff perceives the critical pathway to be, the more successful will be the implementation. Often, vital signs and standard assessments can be documented right on the critical pathway. Critical pathways may be designed to function as a form of charting by exception. If all essential interventions and patient outcomes are included on the critical pathway and variances are routinely addressed, a lengthy progress note is not required. In addition, the critical pathway can be designed to replace existing quality improvement monitors. When this is the case, the quality improvement data are recorded as staff do their documentation.

The multidisciplinary team must also decide who will document on the critical pathway. The critical pathway can be set up to accommodate the team's preferences. In some institutions, each discipline is required to document on the critical pathway any information that is relevant to that discipline. In other institutions, only nursing documents on the critical pathway. When critical pathways are set up in a way that makes them truly useful to each discipline, all will find the pathway relevant for documentation. It is less desirable to require one discipline to do all documentation, because this sets up the pathway to appear as a burden for that discipline, rather than a worthwhile tool.

Critical pathways can be designed so that physicians and other disciplines can order right from the pathway. When the critical pathway is on an integrated computer system, the order immediately goes to the appropriate department, that is, pharmacy, dietary, or physical medicine. This eliminates the lag time that exists when the orders are handwritten and then must be picked up and sent or called into another department.

Critical Pathways and Data Collection

For the critical pathway to be utilized as a quality improvement monitor, the multidisciplinary group decides in advance what kinds of data need to be collected. This can be determined by considering significant processes, care elements, and fundamental outcome criteria. The team also considers what kinds of data will be collected from the critical pathway, as opposed to other sources. Because the potential for data collection is practically limitless, careful consideration is given to the types of data that will be useful for documenting compliance with the critical pathway, achievement of patient outcomes, bench marking, and cost analysis. Much of the compliance and patient outcome data can be documented right on the critical pathway, when designed with this purpose in mind.

Benchmarking and cost analysis are usually achieved from analyzing critical pathway data, as well as comparative data over a period of time. For example, benchmarking might be done by comparing achievement of specific outcomes in one institution with achievement of the same outcomes in comparable institutions. In the early stages of critical pathway implementation, compliance with the critical pathway is a primary focus. However, the focal point of data collection changes over time. Generally, as staff are familiarized with the critical pathway, strict compliance with the critical pathway becomes less of an issue and the focus shifts to achievement of patient outcomes, bench marking, and the cost of delivering care.

IMPLEMENTING THE CRITICAL PATHWAY

Staff Education and Preparation

Before the critical pathway is implemented, staff from each discipline are educated on how to use critical pathways. Depending on staff involvement in the developmental process, the content and the amount of education necessary varies. Staff education for those unfamiliar with critical pathways includes introducing the concept of critical pathways, a comprehensive explanation of advantages, how they will be used and how they will affect documentation and practice. For staff already familiar with the concept, the logistics of incorporating critical pathways into practice on a day-to-day basis is the focus. The education for each discipline is done (at least in part) by a representative from that discipline so that the material is presented in a relevant way. Proper preparation through education of all staff makes the implementation of critical pathways go more smoothly.

Once the critical pathway and associated tools are finalized and the staff are educated in proper usage, it is time to implement the critical pathway. Implementing a single critical pathway initially allows the staff to learn the concept and to adapt to associated changes in a more controlled and less stressful manner. The lessons learned from the "pilot" critical pathway may then be applied to other critical pathways. It is important to make all staff aware of the start date, so that the initiation goes smoothly. Recognition of the initiation of the critical pathway and conveying its significance to the staff can be achieved by marking the day as special. This can be done by having an "unveiling ceremony" where the staff are invited to participate in formally recognizing the implementation of the critical pathway. Any way in which the administrative department of the institution acknowledges the critical pathway and the staff's role in its successful implementation will heighten staff awareness and emphasize the importance of the staff's role in this crucial stage.

Having an individual in charge of overseeing implementation of the critical pathway on the unit or clinical level helps with compliance. This could be the case manager, the clinical specialist for that area, or any staff person who has a solid working knowledge of practice issues and of critical pathways. Being available and working with the staff helps the staff become familiar with the critical pathway and how it works. Without a resource person, questions go unanswered and the staff may either not utilize the critical pathway or employ it inconsistently. When the critical pathway is not used the way it was designed to be used, it is seen as a liability and the benefits are not realized. Minimally, feedback is given at staff meetings that provides specific examples of proper and improper use of the critical pathway.

If a number of staff are making similar errors, these may be addressed with the group. If selected individuals are making errors, these errors are shared with that individual as well as strategies to avoid them. Continuous, consistent feedback, both affirmative and corrective, is the key to compliance and successful implementation of the critical pathway.

From the beginning, the staff are encouraged to incorporate the critical pathway into their practice by using the pathway to organize intershift reporting, interdisciplinary rounds, progress notes, and admission and discharge summaries. The critical pathway provides an organized, multidisciplinary format that keeps staff focused on what is most important for their patients. Because critical pathways are designed around patient outcomes, if all aspects of the critical pathway are addressed, so too are the patient outcomes. The staff need continuous feedback on the use of the critical pathway as a tool to meet their own needs and the needs of their patients.

Dealing with Variances

Once the critical pathway is implemented, and is being used correctly, data begin to accumulate on variances from the critical pathway. When compiled properly, these data provide information, both long term and short term, about how care was delivered, the effectiveness of specific treatments, the legitimacy of the critical pathway, cost information, and achievement of patient outcomes. Variances can be broken down most simply as follows:

- **Patient Variances:** Variances that occur as a direct result of something the patient did or did not do. If the patient does not show up for a treatment or becomes acutely ill and is unable to participate in therapy or to obtain a test or treatment, it is recorded as a patient variance.
- **Caregiver Variances:** Variances that occur when a caregiver is unable to provide care as it is outlined on the critical pathway. If a consultation, treatment, or test is not completed because the caregiver was unable or unavailable to provide the treatment, it is recorded as a caregiver variance.
- **System Variances:** Variances that occur due to a problem in the institution or health care system. An example of a system variance is failure to do a treatment, test, or procedure due to department closure, short staffing, or faulty equipment. Data categorized in this manner is easy to record and compile. In a short period of time, variance patterns will become evident and can be addressed appropriately.

Variances can be further classified into subcategories to address issues specific to an organization or agency. Including variance classifications on the critical pathway forces the staff to recognize variances as they occur. Paying careful attention to variances allows the staff to take a proactive stance and prevent problems. If a treatment is missed and the variance is noted promptly, the patient can be rescheduled for the treatment, before the time delay has an effect on patient outcome. Variances that occur, but do not have a negative effect on patient outcomes are worth noting, but do not require intervention. For example, if a patient is scheduled to start physical therapy on postoperative day 2, but the therapy was started on postoperative day 1, due to rapid patient progress, corrective intervention is not warranted. This type of variance in isolation is meaningless, but if it became a pattern, the critical pathway could be changed to reflect actual patient progress. (For a more in-depth review of variances, see chapter 4).

As compliance with the critical pathway increases, the data retrieved from the critical pathway are used to look at practice and its effect on patient outcomes. For example, if compliance with the critical pathway is good, but patients are not achieving the desired outcomes as outlined on the critical pathway, the staff can zero in on specific parts of the critical pathway to improve patient outcome. If the first critical pathway is used as a pilot study, the multidisciplinary task force should determine in advance the length of time the pilot critical pathway will be in place before moving to finalize the critical pathway and to begin implementation of additional critical pathways. Generally, this decision is based on the number of patients that are admitted under a given diagnosis or procedure. Sufficient information needs to be available to document the merit of the critical pathway, as well as to make necessary changes. The pilot study also gives the staff a chance to accept the practice changes that accompany implementation of the critical pathway. Although the length of time varies from institution to institution, usually one quarter (three months) is sufficient time to gather appropriate data and to give the staff a chance to give feedback and to begin to incorporate the critical pathway into their practice. It is not advisable to allow less than a month, because this does not provide sufficient time for staff feedback and buy-in. Care must also be taken to avoid dragging out the pilot study, because it begins to lose significance and the staff may not take it seriously. After the pilot study is completed, changes are made to the critical pathway as rapidly as possible and the pathway is implemented in full force. New critical pathways can then be added on a regularly scheduled basis.

Data Interpretation

Data retrieved from critical pathway documentation can be used to look at multidisciplinary compliance and practice patterns and their effects on patient outcomes. For example, if compliance is a problem for certain professionals, data are tracked over a period of time comparing outcomes achieved by patients on the critical pathway with data from patients that were not maintained on the critical pathway. If the outcomes of the patients that were not put on the critical pathway are achieved less often than those on the pathway, this data may be used to urge compliance. If the patients that are not maintained on the critical pathway are achieving outcomes with greater consistency than their critical pathway counterparts, these practice patterns are examined along with those outlined on the critical pathway. Perhaps changes can be made to improve the critical pathway, and to promote acceptance of the pathway for previously resistant practitioners.

In addition to supplying data about practitioners, the critical pathway also highlights specific problem areas within the institution that have a negative effect on patient outcomes. For example, staff can zero in on specific elements of the critical pathway to improve patient care. As you may recall, critical pathways are designed to concentrate nursing care and other essential patient services so that the patient can be discharged in a shorter period of time. This may require taking a serious look at the schedule of services provided by the institution. If tests or services are only offered on certain days of the week or at particular times, this may affect the achievement of outcomes and increase the length of stay. Other services, such as physical therapy, occupational therapy, and social work may need to be offered on a more frequent basis, so that the patient gets the required number of consultations or treatments to achieve the stated outcomes in the time period outlined on the critical pathway. Agencies or departments that have traditionally closed down at night, on weekends, and holidays may need to redefine how they do business, so they can meet the needs of their customers.

CONCLUSION

Critical pathways reflect the ideal course of treatment for the average patient admitted under a given diagnosis or procedure. In reality, not all patients follow the critical pathway without deviation. Patients are individuals with unique needs and experiences that must be considered during treatment. The idea behind the critical pathway is not to create a rigid, cookbook way of dealing with a diagnosis. Rather, the critical pathway is a suggested plan of care that includes all of the critical pieces that must be addressed before the patient is discharged from an institution or service. By organizing these pieces, it is hoped that all facets of the patient's illness are addressed before discharge and adequate plans are made for the transition from one place along the health care continuum to the next. Often health care professionals wish they had more time to adequately prepare the patient for discharge, especially in these days of very short lengths of stay. From this venue, critical pathways are a useful, multidisciplinary tool and provide a way of assuring that the patient gets all of the components of care necessary for a smooth discharge and transition to his or her home. With critical pathways, all disciplines work together, in the little time available, to address the patient's needs.

REFERENCES

Coffey, R., Richard, J., Remmert, C., LeRoy, S., Schoville, R., & Baldwin, P. (1992). An introduction to critical paths. *Quality Management in Health Care, 1*(1), 45–54.

Zander, K. (1988). Nursing case management: Strategic management of cost and quality outcomes. *Journal of Nursing Administration, 18*(5), 23–30.

BIBLIOGRAPHY

Alba, T., Souders, J., & McGhee, G. (1994). How hospitals can use internal benchmark data to create effective managed care arrangements. *Top Health Care Finance, 21*(1), 51–64.

Bueno, M., & Hwang, R. (1993). Understanding variances in hospital stay. *Nursing Management, 24*(11), 51–57.

Camp, R., & Tweet, A. (1994). Benchmarking applied to health care. *Joint Commission Journal on Quality Improvement, 20*(5), 229–238.

Campbell, A. B. (1994). Benchmarking: A performance intervention tool. *Joint Commission Journal on Quality Improvement, 20*(5), 225–228.

Denied, R. (1996). Data capture for quality management nursing opportunity. *Computers in Nursing, 14*(1), 39–44.

Dijerome, L. (1992). The nursing case management computerized system: Meeting the challenge of health care delivery through technology. *Computers in Nursing, 10*(6), 250–257.

Dziuban, S., McIlduff, J., Miller, S., & Dal Col, R. (1994). How a New York cardiac surgery program uses outcomes data. *Annuals of Thoracic Surgery, 58,* 1871–1876.

Gottlieb, L., Sokol, H., Oates, K., & Schoenbaum, S. (1992). Algorithm-based clinical quality improvement. *HMO Practice, 6*(1), 5–12.

Hoyer, R. (1995). Prospective payment for home care. *Caring, 14* (3), 28–35.

Julian, K., & Poorer, C. (1991). Nursing case management: Critical pathways to desirable outcomes. *Nursing Management, 22*(3), 52–55.

Kaiser benchmarking study to identify best practices (1994). *OR Manager, 10*(5), 1, 16–17.

King, M., McDonald, B., & Good, D. (1995). Redesigning care using total quality management and outcome/variance analysis. *Aspen's Advisor for Nurse Executives, 10*(5), 3–6.

Krivenko, C., & Chodroff, C. (1994). The analysis of clinical outcomes: Getting started in benchmarking. *Joint Commission Journal on Quality Improvement, 20*(5), 260–266.

Larrabee, J. (1994). Using research to improve quality. *Nursing Quality Connection, 4*(3), 5.

Lenz, S. (1994). Benchmarking: Finding ways to improve. *Joint Commission Journal on Quality Improvement, 20*(5), 250–259.

Lumsdon, K., & Hagland, M. (1993). Mapping care. *Hospital and Health Networks, 67*(10), 34–40.

Mendenhall, S., & Prock, S. (1995). Variance program melds hospital's, vendor's ideas. *Hospital Case Management, 3*(1), 14–18.

Mosel, D., & Gift, B. Collaborative benchmarking in health care. *Joint Commission Journal on Quality Improvement, 20*(5), 239–249.

Nadzam, D. (1994). The indicator measurement system: Change for the better. *Nursing Quality Connection, 4*(3), 6.

Norman, L. (1995). Computer-assisted quality improvement in an ambulatory care setting: A follow-up report. *Joint Commission Journal on Quality Improvement, 21*(3), 116–130.

Nugent, W., & Schults, W. (1994). Playing by the numbers: How collecting outcomes data changed my life. *Annuals of Thoracic Surgery, 58,* 1866–1870.

Saul, L. (1995). Developing critical pathways: A practical guide. *Heartbeat, 5*(3), 1–10.

Schriefer, J. (1995). Managing critical pathway variances. *Quality Management in Health Care, 3*(2), 30–42.

Shikiar, M., & Warner, P. (1994). Selecting financial indices to measure critical path outcomes. *Nursing Management, 25*(9), 58–60.

Simkin, B. (1995). Transitional pathway encompasses outpatient settings. *Hospital Case Management, 3*(1), 1–12.

Tidwell, S. (1993). A graphic tool for tracking variance and comorbidities in cardiac surgery case management. *Progress in Cardiovascular Nursing, 8*(2), 6–19.

Windle, P., & Houston, S. (1995). COMIT improving patient outcomes. *Nursing Management, 26*(8), 64AA–64II.

Woodyard, L., & Sheetz, J. (1993). Critical pathway patient outcomes: The missing standard. *Journal of Nursing Care Quality, 8*(1), 51–57.

3

Working Out Comorbidities and Variances with Copathways and Algorithms

Patricia C. Dykes

The complexity of most health problems necessitates a vehicle for addressing comorbidity when implementing a critical pathway model of health care delivery. Because very few patients present their physicians with a single diagnosis or problem, difficulties occur if comorbidity is not planned for in advance. Critical pathways need to be concise and general so that they can be useful across diverse populations. If strategies are included to deal with common variances within the critical pathway, often the pathway becomes too lengthy and impractical for all patients. Requiring the staff to write separate care plans for each comorbidity is time consuming and unrealistic in most settings. Thus it is beneficial to have a documented strategy for dealing with comorbidities and variances that occur often. Copathways and algorithms have been created to address comorbidity and variance trends, respectively.

COPATHWAYS

Copathways or copaths are defined as an organized plan to address the essential patient outcomes that must be considered for a particular comorbidity or problem over a given length of stay. Copaths are specifically designed to address comorbidity or problems that result in patient variance in a clear, concise, and patient-focused manner. Depending on the format, copaths may be employed as a multidisciplinary tool, or they may be devised to replace a portion of the nursing care plan.

Some institutions have tried to resolve the problems created by comorbidity by putting patients on more than one critical pathway. This often creates an additional burden on the staff and is in many cases redundant. A more workable solution is to create copathways to address common comorbidities. These copathways are used across departments within institutions or agencies, because many comorbidities are not unique to medicine, surgery, or psychiatry but are seen across settings. Rather than mapping out each intervention for the patient over a given length of stay as critical pathways do, copathways simply outline the patient outcomes that must be met before the patient is discharged. The staff documents each outcome, as it is met, directly on the copathway. Any outcomes that are not met at the time of discharge must be addressed in the discharge summary.

An example of a copathway is the Homicidality/Suicidality Copathway (Fig. 3.1). This can take the place of a separate nursing care plan written to address the related patient safety issues. Essential patient outcomes are listed in the left-hand column, and the nurse has the option of signing the outcome off as "met" or initialing the "not met" column and addressing what needs to be reinforced or taught in the "comments" column. As with a nursing care plan, the copathway can be individualized for each patient and clearly shows what patient outcomes have been met and what outcomes still need to be addressed before discharge.

Copathway variances, like critical pathway variances, are classified as patient related, caregiver related, or system related. In this way, the copaths are used in conjunction with the critical pathway as part of the continuous quality improvement process.

ALGORITHMS

As mentioned in earlier chapters, when critical pathways are implemented properly, they utilize the principles of continuous quality improvement (CQI) and total quality management (TQM). Vigilant monitoring and data analysis will highlight variance trends frequently associated with specific diagnoses and procedures. The next logical step is to take the results of the data analysis and feed it back into the system to make it more efficient. One method of using data to improve the system is to create algorithms that address common variances, with the hope of preventing future variances.

Algorithms are defined as "written guides to stepwise evaluation and management strategies that require observations to be made, decisions to be considered, and actions to be taken" (Hadorn, 1992). Algorithms have been utilized for years as teaching tools for residents and other professionals in training to help facilitate decision making. As with critical pathways, algorithms are meant to be used as a guide and never as a substitute for professional clinical judgment and decision making. The purpose of using algorithms in conjunction with critical pathways is

Patient Name: _____

MR #: _____

High Risk for Violence to Self or Others Related to:

☐ History of suicide attempts ☐ Provocative behavior   ☐ Feelings of unreality
☐ Depressed mood   ☐ Low self-esteem   ☐ Rage reactions
☐ Use of suicidal gestures for manipulation of others   ☐ Other_____

As Evidenced By:_____

LONG-TERM GOALS:		
☐ Patient will not harm self or others while in the hospital.		
☐ Other:		.

SHORT-TERM GOALS:		
☐ Patient will seek out staff member if feelings of harming self or others occur during the shift.		
☐ Other:		.

PATIENT OUTCOMES:	Date/Initials MET	NOT MET	*COMMENTS:
1 Verbalizes ability to refrain from acting on suicidal impulse at this time.			
2 Identifies 3 precipitating factors to feelings of suicidality.			
3 Identifies behaviors that indicate anxiety is increasing and ways to intervene before violence occurs.			
4 Identifies and practices one physical outlet for anger and anxiety, such as exercise or yoga.			
5 Identifies one (1) major stressor and verbalizes one (1) positive coping mechanism that can be applied to that stressor.			
6 Patient/significant other identify three (3) resources to call for assistance or support when feeling suicidal and personal coping strategies are unsuccessful.			
7 Identifies the purpose of the Suicide Hotline.			
8 Verbalizes a willingness to utilize the Suicide Hotline if needed.			
9 Other:			

*If not met or N/A, please comment

Outcome Codes: (check appropriate column)	Variance Codes:	Initials/Signature:
M=Met	P=Patient Related	_____
NM=Not Met (Must Comment)	C=Caregiver Related	_____
N/A=Not Applicable (Please Comment)	S=System Related	_____

FIGURE 3.1 Homicidality/suicidality copathway.

to address formally common variances and to standardize variations in practice (Schriefer, 1994).

There is a paucity of literature on the use of algorithms in critical pathways because critical pathways are still in their infancy. The first work linking critical pathways with algorithms as a tool for addressing common variances has been documented by Schriefer and her colleagues at the Medical Center of Vermont (1994).

Schriefer described algorithms as tools that may be utilized to assist clinicians with decision making within a plan of care (such as a critical pathway).

Literature and practice standards assist in creating algorithms, and this allows practitioners to address high variance areas and complex patient problems. Decision making is usually most complex along the critical pathway at the point that most variances occur (American Health Consultants, 1995, p. 21). Algorithms assist the practitioner in the problem-solving process and in making prudent decisions regarding care. Although the benefits of combining critical pathways and algorithms is just beginning to be studied, data available to date suggest that their use is associated with a reduction in variances and improved patient outcomes (Schriefer, 1994).

Creating Algorithms

The need for algorithms is often initially highlighted by variance analysis data. Therefore, they are generally not put into place until after the critical pathway has been in use for a period of time. One exception would be the identification of a problem or significant practice variability noted during the initial chart review. In this case, an algorithm would be written to address the problem and implemented along with the critical pathway. Generally, once the critical pathway is in place, if there is not improvement noted in patient outcomes, length of stay, or cost containment, it is possible that an examination of common variances may reveal the need for an algorithm. The flow chart in Figure 3.2 outlines the decision-making process involved in the creation of a copathway or an algorithm.

Algorithm Development

The following examples illustrate algorithm development:

- **Example #1**

  Schriefer and colleagues at the Medical Center Hospital of Vermont were able to demonstrate the potential for combining critical pathways and algorithms for affecting patient outcomes. Monthly review of variances from the CABG (Coronary Artery Bypass Graph) critical pathway revealed a shortage of beds in the surgical intensive care unit, which was having an impact on the number of surgical cases that could be done. As a result, a significant number of cases were consistently being canceled. By gathering the team of players involved in these cases to study the variance, and by reviewing current literature and practice standards, an early extubation algorithm was created. After implementation of the early extubation algorithm in the surgical intensive care unit, Schriefer and colleagues documented a decrease in SICU (Surgical Intensive Care Unit) readmission rate, an overall reduction in the length of stay by 1 day, and a reduction in the annual cost of CABG surgery by $3.5 million (Schriefer, 1994).

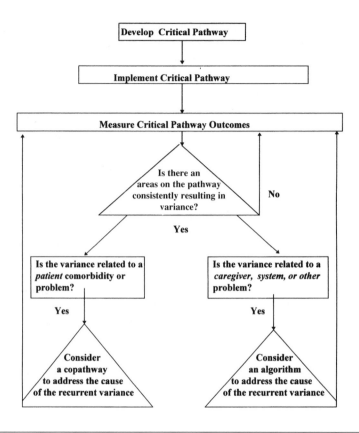

FIGURE 3.2   Combining copathways and algorithms with critical pathways.

- **Example #2**

   At Danbury Hospital in Danbury Connecticut, an antibiotic algorithm was developed to address practice variability for patients admitted for treatment of pneumonia. Regular review of variances and retrospective chart review revealed a wide variation in the types of antibiotics ordered to treat simple pneumonia. It was found that powerful and expensive antibiotics were routinely ordered and continued for the patient, even if the culture later revealed sensitivity to less powerful, less costly antibiotics. Financial concerns, coupled with concerns about superinfection, propelled the task force to study the issue in more depth. Further study revealed that the patients admitted under the diagnosis of simple pneumonia (DRG 89) could generally be broken down into two categories, those from the community who acquired the pneumonia at home, and those who acquired the pneumonia in a long-term care or other institutional setting. Patients that acquired pneumonia at home were more likely to have a less drug resistant strain of bacteria as the cause of the pneumonia than the patients from an institutional setting. Because patients are generally started on an antibiotic before the final culture and sensitivity are available, an algorithm was developed to guide physicians through this process (see Figure 3.3). In addition to the initial cost savings realized for both the patient and the institution, the antibiotic algorithm has the potential for promoting responsible antibiotic usage that discourages future drug resistance.

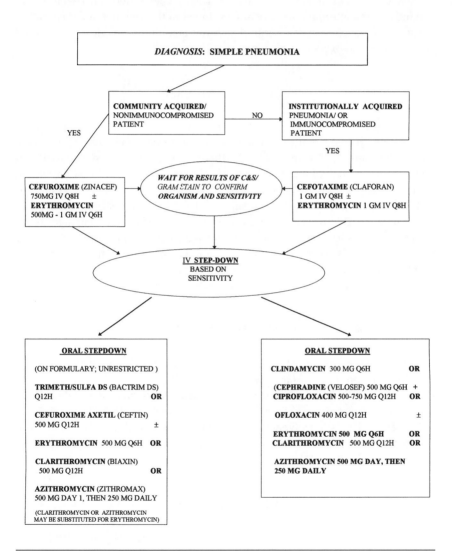

FIGURE 3.3 Simple pneumonia: Antibiotic recommendations algorithm.

Note. Adapted from "Antibiotic Recommendations," by R. Hindes M.D., Danbury Hospital, Danbury CT.

Developing Algorithms

The process of developing algorithms utilizes the same principles as those used for developing critical pathways. Although algorithms may be borrowed from similar institutions or agencies for reference, ultimately they should be developed for use by a multidisciplinary team involved with the care of the patient the algorithm addresses. Content should reflect current practice standards, documented research, and industry bench-mark standards. Developed in this manner and then implemented in conjunction with the critical pathway, algorithms provide a practical and concise approach to addressing common variances.

CONCLUSIONS

As critical pathways evolve, attention needs to be directed toward developing ways of dealing with comorbidity and common variance trends in a manner consistent with a CQI and TQM format. Ideally, this needs to be accomplished without imposing the burden of additional documentation on the staff. Copathways and algorithms are tools that address all classes of variances logically and systematically while augmenting the critical pathway. This is accomplished without creating an unreasonable workload for the staff. Without copathways and algorithms, critical pathways lose their applicability to large patient populations and are only useful for the fraction of patients that present the physician with a single health problem. As a result, the future of critical pathways is assured significance only when linked with copathways and algorithms.

REFERENCES

American Health Consultants (1995). Early extubation reduces surgery cancellations. *Hospital Case Management, 3*(2), 21–25.

Hadorn, D. C. (1992). An annotated algorithm approach to clinical guideline development. *Journal of the American Medical Association, 267*(24), 3311–3314.

Schriefer, J. (1994). The synergy of pathways and algorithms: Two tools work better than one. *Joint Commission on Accreditation of Healthcare Organizations, 20*(9), 485–499.

BIBLIOGRAPHY

Gottlieb, L., Sokol, H., Oates, K., & Schoenbaum, S. (1992). Algorithm-based clinical quality improvement. *HMO Practice, 6*(1), 5–12.

Niemeyer, S. (1995). Policy on practice guideline development. *Home Health Digest, 1*(9), 2–8.

4

Data Collection, Outcomes Measurement, and Variance Analysis

Patricia C. Dykes
Debra A. Slye

CRITICAL PATHWAY DESIGN AND DATA COLLECTION

An important component of designing the critical pathway is considering the types of data that will be collected once the critical pathway is in place. When set up properly, critical pathways have the potential for being powerful data collection tools. The data retrieved from the critical pathway may be used for documenting compliance with the critical pathway, achievement of predetermined goals and outcomes, and can become part of a database used for research and internal benchmarking.

An integral part of this process is working out in advance the types of data that will be collected from the critical pathway, the purpose of the data, who will document on the critical pathway, and who will do the actual data collection. The central question becomes, how do you integrate patient care charting and data collection without creating unreasonable amounts of paperwork?

Types of Data

When deciding what types of data will be collected using the critical pathway, it is helpful to consider the purpose the data will serve for the institution. There are five main uses of clinical outcome data documented in the literature (Docherty & Dewan, 1995). These include the following:

- administration and clinical operations management
- compliance with regulatory agencies
- benchmarking
- marketing
- research

Data from the critical pathway can be used to serve each of these purposes. Administrative and operations management refers to the way clinical practice and achievement of outcomes are monitored in an institution. By keeping track of this type of data, administrators are able to document the effectiveness of the care that is provided. In turn, this data may be used for continuous quality improvement, internal benchmarking, or competing for business in the managed care arena.

Administrative and operations management data may also be used to address systems issues and the relationship of these issues to patient outcomes. For example, systems variables such as staffing patterns, levels of providers, or technology and treatments used for patients under a particular diagnosis can be examined in terms of their effect on the achievement of patient outcomes (Docherty & Dewan, 1995). Over time, the effect on cost can also be calculated. Administrative and operations management data have the potential for influencing significant system changes.

In addition, the effect of variances from the critical pathway on patient outcomes may also be collected under administrative and operations management data. The resulting data have the potential for changing practice over time. Until now, the relationship of providing care in a certain way and achievement of outcomes has never been looked at scientifically. As more health care providers become computerized and have the potential for collecting and analyzing large amounts of data, the implications for practice are consequential.

Documentation and Data Collection

Several important questions regarding data collection need to be addressed when designing the critical pathway.

(1) Will the critical pathway be a documentation tool or simply a guide?
(2) If the critical pathway is a documentation tool, what disciplines will chart or record variances on the critical pathway?
(3) What types of data will be collected?
(4) How will the variance data be collected and how will this data be analyzed?

When critical pathways initially were being trialed, often they were simply used as practice guidelines to cue each discipline into key interventions. Over time, this method has proved to be impractical because when critical pathways are designed as guides, staff often consider them "optional." On slow days, the staff might refer to the critical pathway and chart variances, but when acuity is high, critical paths represent one more "unnecessary piece of paperwork" that will not get done.

Ideally, all disciplines should document on the critical pathway. Requiring one discipline to shoulder the responsibility for all critical pathway documentation takes what is designed to be a "multidisciplinary" tool and renders it unidisciplinary. In addition, when one discipline is responsible for all documentation, it is compelled to take on a "policing" function, monitoring the compliance of other disciplines.

When critical pathways are designed using multidisciplinary input, they can be utilized by each discipline for documentation and for report. If the critical pathway truly contains all of the care elements and interventions necessary for the majority of patients admitted under a given diagnosis, all disciplines should be able to use the critical pathway as the basis of their documentation.

Some institutions use case managers, separate and distinct from staff nurses, to complete the nursing documentation on the critical pathway. Critical pathways provide a sound tool for case managers to track patient outcomes. Unfortunately, many case managers are following too many patients to be able to address realistically each variance *as it occurs* with every patient. If variances cannot be addressed in a timely manner, variance collection does not help the patient on the critical pathway, only *future* patients. Much of the documentation on the critical pathway then becomes an exercise in futility.

Ideally, the staff nurse should also document on the critical pathway. As the day-to-day manager of patient care, the staff nurse is in an optimal position to recognize variances in a timely fashion and to act on them. Because of the amount of interaction the staff nurse has with the patient, he or she is able to influence heavily patient outcomes over the course of a day. The staff nurse needs to be acutely aware of both the patient's progress along the critical pathway and interventions that promote the achievement of desirable outcomes. Documentation on the critical pathway brings variances to the staff nurse's attention, so that they may be addressed. Using the critical pathway as a reporting tool allows the staff nurse to follow through on variances that require ongoing vigilance.

As health care organizations continue to downsize, the role of the staff nurse will become more heavily dependent on the professional role functions of the nurse (such as case manager), and the nonprofessional roles and tasks will be delegated to lesser skilled workers. The days when an institution can afford to pay a large group of professionals a salary and then ask them to do a variety of nonskilled tasks are numbered.

OUTCOMES MEASUREMENT

Successful implementation of case management is dependent on the development of reliable tools to measure patient care outcomes (Crummer & Carter,

1993), to analyze variances, and to update pathways in current use (Lumsdom & Hagland, 1993). Most quality management efforts today focus on outcomes improvement—clinical and fiscal. Using data to ensure the deployment of cost-effective processes involves knowing what works (efficacy), using what works (appropriateness), doing well what works (execution), and considering the values that underlie these processes (purpose) (Berwick, 1988). "All four areas must be pursued effectively if health care quality is to be successfully defined, measured, and protected" (Fein, 1995).

The relationship among structure, process, and outcome is shown in Figure 4.1. The process of care is carried out within a structure, the result of which is outcomes. Outcomes can be divided into patient outcomes, care provider outcomes, and health care enterprise outcomes.

Patient Outcomes

Indicators of patient outcome include mortality rate, complications, patient satisfaction, and improved quality of life as evidenced by increased longevity and improved functional capacity (self-care abilities, symptom management, health-promoting behaviors). Morbidity and mortality rates are the most commonly measured indicators of patient outcome. However, measurement of patient satisfaction is gaining priority. Measurement of functional capacity pre- and post-treatment is essential to determine whether the health care process did the patient any good.

Questions that need to be addressed include:

(1) Is the patient better off as a result of this health care experience?
(2) Is the quality of life improved, maintained, or has it declined?
(3) Is the patient able to care for himself or herself, manage his or her own symptoms, and demonstrate health-promoting behaviors?

Outcome models such as the Medical Outcomes Study 36-Item Short Form (SF36) and the Functional Independence Measure (FIM) are available and in the public domain, whereby care providers can measure patients' functionality at various stages throughout their disease management and recovery to determine the value of the care received. Global measures of outcomes assessment do not replace the need for clear definitions and measurement of the patient's response to treatment, taking into account the severity of illness and the presence of comorbidities (Tallon, 1995).

Care Provider Outcomes

Care provider outcomes include increased satisfaction with the care delivered, autonomy in practice, participation in decision making, and reduced turnover rates.

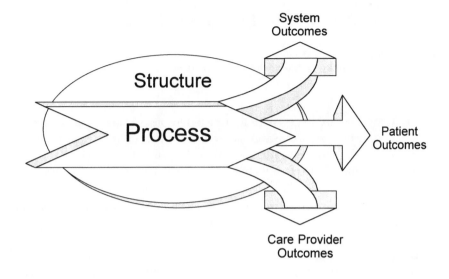

FIGURE 4.1 Quality improvement components.

Studies have shown that positive care provider outcomes lead to better patient outcomes. In a study of nine critical care units, improved patient outcomes (as evidenced by reduced risk-adjusted mortality) were demonstrated in units with the following (Zimmerman, Shortell, & Rousseau, 1993):

- a patient-centered culture
- strong medical and nursing leadership as evidenced by shared visions, supportive visible leaders, and a collaborative approach to problem solving
- effective communication and empowered nursing staff

Health Care Enterprise Outcomes

Outcomes related to the health care enterprise are primarily related to costs. Length of stay is commonly correlated with cost; however, some studies show that the majority of costs can be incurred within the first few days of a patient's stay.

Pathways are not always designed to focus on the outcomes of the care delivered (i.e., the patient's response), but rather on the documentation of what the care provider did. From a quality perspective, it is most important to document what the patient learned as opposed to what content was taught. When designed from a patient outcome perspective, the evaluation of progress along the pathway becomes a verification of outcomes (Adams & Biggerstaff, 1995).

ANALYZING THE CARE PROCESS

Analysis of pathways, variances, and outcomes on a cross-patient basis helps answer critical questions facing health care providers today.
   For example:

   (1) What is the effectiveness of particular interventions or pathways on outcomes?
   (2) What are the costs of care associated with a particular pathway or intervention?
   (3) What is the frequency of use of a particular pathway?
   (4) What is the effectiveness of an individual practitioner's interventions?

   The ability to answer these questions is becoming increasingly a priority in the heavily regulated, dominant, health care atmosphere. Consistent critical pathway data collection and analysis will quickly become the standard for organizations seeking accreditation and regulatory compliance. Once a solid continuous quality improvement program is in place, it can be built upon to meet the needs of the organization as well as the accrediting and regulatory agencies. The primary goal of accrediting and regulatory agencies is to warrant that the needs of the consumer are expertly and consistently met. A solid CQI program, based on data retrieved from critical pathways, incorporates this goal into the vital functions of the institution.

Regulatory Agencies and Data Analysis

Currently, many states are considering legislation that would require hospitals to collect and then submit outcome data to a central agency. In 1988, Pennsylvania passed such a law and created a new state agency to analyze outcome data to determine the efficacy of hospital care. Since 1992, this agency, the Health Care Cost Containment Council, has been publishing physician and hospital-specific information available to consumers who are selecting health care providers (Krivenko & Chodroff, 1994). In New York State, a risk-adjusted outcome system is used to monitor all cardiac surgeries. The results of this survey are released to the public and have had profound implications for patients and providers (Dziuban, McIlduff, Miller, & Dal Cor, 1994).

   In addition, the Joint Commission on the Accreditation of Healthcare Organizations (JCAHO) has initiated a voluntary sharing of outcomes data as part of its Indicator Measurement System (IMSystem). The IMSystem is a national performance measurement system that is designed to provide participants with comparative information to be used ultimately to improve standards of care nationally. Currently, participation in the IMSystem is voluntary, but it will become a requirement for accreditation before the year 2000 (Nadzam, 1994).

Outcome Data and Benchmarking

Clinical outcome data from critical pathways may also be utilized for marketing services and for bench marking. This data will provide a tool for documenting excellence where practice exceeds the market standard. Critical pathway outcome data can be used to promote services and to serve as a guarantee of the care provided. This same data can then be employed to compare treatment outcomes with other providers. Open exchange of outcome data will improve standards of care across the health care industry, as all providers continuously work to improve their outcomes in a competitive marketplace. Once sufficient outcome data are available, bench marking can be done with leaders inside and outside the health care industry (Mosel & Gift, 1994). External bench marking (called competitive and functional bench marking) allows members in organizations to learn from leaders in health care and other industries that have perfected both clinical and systems issues.

Kaiser, a well-known health maintenance organization (HMO), describes benchmarking as "a process of identifying best practice, defined as the place where service, cost and quality meet" (Patterson, 1994, p. 16). The introduction of competition into the health care marketplace has necessitated this continuous process of measuring and comparing an institution's business processes against industry leaders, with the objective of improving performance (Patterson, 1994).

Resource Utilization

In addition, clinical outcome data may be used to develop research-based practice, otherwise known as research utilization (Larrabee, 1994). When used as a data collection tool, critical pathways provide a large quantity of practice data that can subsequently be analyzed to determine the effectiveness of key interventions or outcomes. With the assistance of a computerized system, administrators in institutions have the capability of analyzing and documenting systemwide outcomes as well as the treatments and performances of individual practitioners.

"The *clinical* pathway can be used as a cost and resource allocation tool, the resolution of problems as evidenced by the achievement of goals and expected outcomes tell you whether it's working and the variance analysis tools provide the foundation for your quality improvement database" (Lumsdon & Hagland, 1993). A variety of tools to estimate resource utilization are currently available in the market. However, their completion is commonly a secondary process to care documentation and, therefore, often not clearly reflective of the actual care delivered.

Ideally, resource consumption is automatically acquired as a direct result of care documentation so that it reflects most accurately the actual care delivered. Tying a relative value to an intervention multiplied by the level of expertise of the provider delivering the care is one mechanism of defining resource utilization. Calculating resource consumption by pathway supports forecasting of resource requirements. Analysis of variances in resource consumption and associated outcomes by pathway may facilitate refinement of the pathway to optimize the cost-effectiveness of care.

VARIANCE ANALYSIS

Simply stated, a variance may be defined as the difference between how patient care and outcomes were defined on the critical pathway and what actually happened. Timely recognition of variances is an essential step toward assuring success of critical pathways and delivery of quality patient care. The intrinsic value in developing standardized pathways is twofold (Tallon, 1995):

- to guide the care for the 80% of the cases that fall within the normal range
- to facilitate early detection of variance from the standard and rapid intervention to get the patient back on track

Analysis of variances is primarily useful to consider which variations in interventions contribute, either positively or negatively, to the patient's achievement of outcomes as predicted. In addition, variance trends can provide useful information for the quality improvement plan. A clinical practice may target certain quality indicators for attention based on those identified in their variation analysis. Any patients that meet the criteria for any of the identified quality indicators may be triggered by a flag so that care may be guided to eliminate or minimize the impact of that indicator (Crummer & Carter, 1993).

When considering and classifying variances, analysis of the care process occurs on two levels:

- for the individual patient
- on a cross-patient basis

Variances can be defined as discrepancies between planned and actual events, outcomes that differ from anticipated, or deviations from the projected time line. Analysis of variance on an individual patient basis helps fine-tune the patient care process to meet the needs of that patient. Cross-patient analysis of variance facilitates measurement of process effectiveness in achieving desired outcomes. Analysis of aggregated patient data is essential to refine standardized pathways, to risk-adjust your data, and to bench mark against regional or national norms.

Based on identified variances, particularly those that may be outcome-bearing, care may need to be adjusted for an individual to get the patient back on track. From a cross-patient basis, variances must be analyzed to determine if changes should be made to the standard pathway to make it more realistic. Variances are commonly classified into those attributable to the patient, the care provider, and the health care enterprise (see Figure 4.2).

Variances attributable to the patient are further critiqued to see if such variations are consistent enough across the patient population to justify the need to change the standard pathway. Care provider variances may be critiqued to determine if performance problems exist, if resources were insufficient to provide the care as defined by the standard pathway, or if there is sufficient deviation from the standard based on individual care provider preferences to warrant changing the standard pathway. Variances related to the health care enterprise are critiqued to determine if barriers exist within the system that prevent patients from progressing according to the standard. A model for variance analysis is shown in Figure 4.3. Refinements to pathways can also be tested using a cross-patient analytic tool. A study group of patients may be assigned to a test pathway. Variances and outcomes experienced as a result of the test pathway can be compared to those observed in patients on the existing standard pathway for the same diagnosis. Such analyses can validate the effectiveness of the test pathway and provide evidence to support its substitution for the existing standard.

Variance Tracking

There are four methods of variance tracking (data collection) documented in the literature. These include: (1) identifying variances retrospectively through chart review, (2) documenting variances on a separate quality assurance (QA) monitor, (3) using computerized systems to track variances and, (4) documenting variances right on the critical pathway (Schriefer, 1995). The first two methods, the retrospective chart review and the separate QA monitor, are inefficient because they are labor intensive and paper oppressive. Although variances that are identified retrospectively through a chart review may be valuable as a CQI tool for future patients, they are often recognized after it is too late to make changes that will have an impact on patient outcomes in the present tense. Charting variances on a separate QA monitor requires the introduction of yet another piece of paperwork in an already laden system.

The ideal method of data collection is the use of a computerized system that is preprogrammed to collect data as multidisciplinary staff documents. Such a format allows for ordering right from the critical pathway, alerts the staff to variances as they occur, and performs complex data analysis. To be truly functional, the documentation system should also be integrated with other system databases such as clinical, utilization review, and financial (Schriefer, 1995). In this way, the system can alert the staff to financial as well as clinical variances. For example, if a physician orders a costly medication, the computer could automatically run a cost analysis and list less costly alternative medications, with the same desired effect. Ultimately, it would be up to the physician to utilize clinical judgment to order the medication that would be most beneficial for the patient.

Patient/significant other variances

- Inability to learn skill needed for self-care at home
- Inadequate social support or systems at home
- Not indicated at this time for patient/family
- Unable to return to preadmission environment
- Patient/family decision
- Complication or condition (physiological/psychological)
- Patient condition warrants early discontinuance
- Patient noncompliance
- Patient/family unavailability
- Other _____

Care provider variances

- Lack of or inadequate documentation
- Physician response time
- Other provider response time
- Physician/provider error
- Time orders were written
- Orders outside clinical pathway parameters
- Treatment or medication omitted
- Not ordered by physician(s)'/physician preference
- Other_____

System variances

- Bed availability
- Schedule conflict
- Consultant unavailable
- OR time unavailable
- Results/data unavailable
- Supply/equipment unavailable
- Department closed
- Placement unavailable
- Home health unavailable
- Pending payer approval
- Other _____

FIGURE 4.2   Variance categories.

Note. Adapted with permission from Saint Joseph Hospital, Denver, CO, 1996.

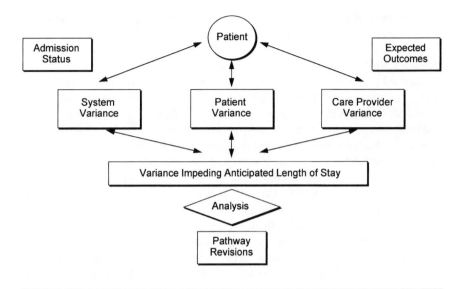

FIGURE 4.3   Variance analysis model.

Note. Adapted with permission from Saint Joseph Hospital, Denver, CO, 1996.

The types of data that will be collected and analyzed will also need to be considered when designing the critical pathway and the variance analysis process. If a computerized critical pathway and documentation system is in place, the quality of data collection is limited only by the system in use, the content of the critical pathway, and subsequently the computerized documentation of the staff. If a paper critical pathway and documentation system are in use, then limitations include staff documentation, critical pathway design, data collection, recording of variances, and data analysis.

Variance Data Collection

Regardless of the type of critical pathway, computerized or paper, compliance data will need to be collected in the beginning stages of implementation. This data will be used to document the effectiveness of the critical pathway, to make critical pathway and practice modifications based on hard data, and to monitor compliance of multidisciplinary staff around the use of the critical pathway. A computerized system will keep track of compliance data and do necessary calculations. A noncomputerized system will need to be set up efficiently so that the data can be scanned into a statistical program, or organized in such a way that the data can be easily extracted and simple calculations may then be done by hand. One way to achieve this is to have a checkoff system, whereby interventions and patient outcomes are checked off as they are completed. If they are not checked off, they are considered a variance and need to be addressed in the progress note. The format of the critical pathway in Table 4.1 allows the staff to check off completed interventions and outcomes and to classify significant variances.

Table 4.1 is an excerpt from an inpatient critical pathway for anxiety. Each intervention is preceded by a box, which is checked off as the intervention is completed. If the intervention is not completed on the day specified, this constitutes a variance and must be addressed in the progress notes. One way to document variances in progress notes is to utilize a Variance-Intervention-Evaluation or V-I-E format (Hronek, 1995). The V or variance section of the note addresses any variance that may have occurred on a given day and the reason for the occurrence. In the I or intervention section of the note, the staff list interventions done to correct the variance. In the E or evaluation section of the note, the staff evaluate the effect that the variance will have on desired patient outcomes, given the interventions that were taken. For example, if the "family meeting scheduled" box is not checked off under day 1, the professional responsible for arranging family meetings (usually the nurse or the social worker) would need to write a V-I-E note. The V-I-E note might read as follows:

V: Unable to schedule family meeting because patient's daughter (Ms. L.) works until 10 p.m. and was unable to return my call.

I: Message left on daughter's answering machine with instructions to call back tonight and talk with patient's primary nurse (Ms. C.) about possible meeting times. This information will be used to schedule a family meeting in a.m.

E: Family meeting will be scheduled in a.m. and should not prolong discharge readiness.

Variance Documentation

Most health care institutions have taken steps toward computerization, but have not achieved this as yet. If the documentation system is not computerized, the next best scenario is documentation right on the critical pathway. Every effort should be made to condense other required documentation so that the critical pathway is utilized to its fullest potential. When designed with this purpose in mind, variances can be recorded on the pathway and addressed in the progress notes. By recording variances on the critical pathway, the staff are made aware of variances as they occur and can take remedial action when necessary to improve patient outcomes.

After a patient is discharged, his or her variance data can be scanned by computer into an ASCII file, and then transferred to a statistical program for correlation and analysis. Data analysis that is done without computer assistance is more time consuming, but proper organization of the critical pathway allows for some

TABLE 4.1 Critical Pathway for Anxiety.

Care Elements	Day 1	Day 2	Day 3	Day/ Discharge Outcomes
Clinical Assessment	☐ *Suicide risk__ ☐ Multidiscipli-nary assess-ment__ ☐ PE complete ☐ Initial family meeting scheduled__	☐ Competency assessed. ☐ Diagnosis confirmed.__ ☐ Formal psych testing (if unable to identify diag-nosis)__	☐ OT/TA assessment completed.__ ☐ Psychosocial completed.__	☐ Assessment for d/c transition readiness. Affect/mood appro-priate. No evidence of thought d/o. ☐ Denies suicidal ideation.

*Presence requires documentation.

simple calculations to document progress. The design of the critical pathway in Table 4.1 can be used for paper documentation and manual data analysis. The format can then be easily transferred to a computerized documentation system, when that becomes a reality. Because the variances are documented in the progress note section of the chart, the documentation system will not need to change with the advent of computerization.

Data Analysis

Once critical pathways are implemented, data from the critical pathway will begin to accumulate and will need to be analyzed. It is a good idea to have a strategy for analyzing this data before actually implementing the critical pathway. Some institutions have a specific person or department that is responsible for data analysis and outcomes management. Other institutions utilize members of the team that designed the critical pathway to address variances and to make appropriate changes. Without data analysis and a formal evaluation of the analyzed data, critical pathways will not function as the CQI tools they are designed to be. It is through the continuous collection of data and review that practice will be substantiated or changed to improve patient outcomes.

Documentation on the critical pathway can be analyzed over a given period of time (for most diagnoses a quarter is sufficient) to show compliance with the critical pathway and achievement of patient outcomes. The results should be analyzed in a timely fashion and made available to all involved with the critical pathway. By making the results available, the staff can see how their interventions influence patient outcomes. In addition to offering incentive and encouragement to the staff consistently using the critical pathway, the results can be used to encourage compliance in noncompliant staff. For example, some simple correlations may be done using the mean, standard deviation, and the range that document the relationship of compliance with the critical pathway and length of stay, complications, and patient satisfaction. This data can be broken down by discipline, cost, achievement of outcomes, or other factors. Noncompliant staff may be able to ignore one or two reports, but consistent, timely reporting of results will get the attention of those reluctant to use the critical pathway. It also provides hard data for supervisors and section chiefs to evaluate performance.

Organizing the Report

When writing up a report, it is important to organize the results of the data analysis in a format that is clear to all parties that will review it. The reports do not need to name individual practitioners when reporting compliance data and patient outcomes. All staff can be assigned a number and the results can be reported according to staff code. The staff can get their code from their supervisor or section chief. Most professionals are familiar with their own practice and the practice of their colleagues. Consistent monitoring and reporting of results sets up a sense of competition within the staff to emulate practice associated with achieving patient outcomes. A sample report form is included in Figure 4.4.

The critical pathway can also be used as a tool to collect long-term data. Listing outcomes right on the critical pathway (see Table 4.1) allows for tracking of achievement of predetermined patient outcomes. This also sets up outcome-based discharge criteria, which are more patient-centered than focused solely on length of stay. Data on comorbidities can also be collected for each diagnosis. This information can be correlated long term with complications. The resultant findings may be used to develop algorithms for subcategories of patients under specific DRGs. For example, if a large group of patients admitted under DRG 210 (fractured hip) with a comorbidity of diabetes had a prolonged length of stay, the group doing the CQI would go back and look at variance data and do chart reviews on these patients to find similarities. They might find that elevated blood sugar levels correlated with prolonged wound healing or with postoperative infection. This information could then be used to create an algorithm to be followed for patients admitted under DRG 210 who also have a secondary diagnosis of diabetes mellitus. Future data collection and analysis would explore the relationship of the use of the algorithm and achievement of patient outcomes such as postoperative complications, morbidity, and length of stay.

Evaluating Cost

Another area that warrants careful attention when analyzing critical pathway data is cost data. Although all of the areas mentioned are important to address with vari-

DRG 210 Quarterly Report
10/93 - 12/93

- **DEMOGRAPHICS:** Organize demographic data to paint a picture of the population that you are addressing. For example: Sixty-two charts were reviewed this quarter for all patients admitted under DRG 210 (fractured hip). The mean age was 76 years with a range of 26 to 96 years. Sixty-eight percent of patients were female amd 32% were male.

- **LENGTH OF STAY (LOS):** Report the current mean length of stay and then compare to previous quarters. For example: The mean LOS was 10.3 days with a range of 3–27 days. This is down 1 day from the last quarter, and 3 days for the year. The trend is as follows:

	1/92– 12/92	1/93– 3/93	4/93 –6/93	7/93– 9/93	10/93– 12/93	Cumulative Total 1993
Mean LOS	19.06	13.04	9.6	13.48	10.3	11.6
N	109	23	25	62	41	151
SD	11.03	6.02	5.91	9.36	5.1	1.8

When the LOS is reported in the above format, one can easily see the downward trend in LOS, as well as a significant decrease in length of stay for the year.

LOS may also be broken down into presurgical and postsurgical LOS to illustrate system issues. In addition, if the sample size is large enough, LOS may be broken down by practitioner to illustrate the effect of different practice patterns on LOS. This type of analysis needs to be documented over time and with large sample sizes before conclusions may be drawn related to significance.

- **PLACEMENT:** Discharge placement is one measure of functional ability at discharge. By looking at placement and then breaking down placement by LOS, you are able to look at functional ability at discharge as well as systems issues. For example, discharge placement may be broken down as follows:

    42.5%: Discharged to home
    42.9%: Discharged to ECF
    14.6%: Discharged to rehabilitation facility

The LOS by placement is as follows*:

	Home	**ECF**	**Rehabilitation**
Mean LOS	12.5	16.8	8.1
N	26	27	8
SD	3.8	12.6	2.8
RANGE	6–21 days	4–48 days	3–21 days

*The data in the previous chart show a prolonged LOS for patients discharged to ECF. This could be studied further, if this proves to be a long-term trend.

FIGURE 4.4 Sample report form.

- **COMPLIANCE ISSUES:** Any issues of noncompliance with the critical pathway may also be addressed in the report. For example, mobilization, physical therapy, removal of foley catheters are all interventions charted on the critical pathway, with a potential of increasing morbidity and length of stay. If compliance with early mobilization is a problem, mobility can be correlated with length of stay as well as discharge placement to show the effect of early mobility on these outcomes. In addition, foley catheter removal may be correlated with incidence of UTI, as well as length of stay.

  If compliance is an issue with specific practitioners, this can also be addressed in the report by correlating functional level, LOS, morbidity, and patient satisfaction with physician.
- **PATIENT SATISFACTION:** Patient satisfaction can be correlated with compliance with the critical pathway, as well as with preexisting data before implementation of the critical pathway.
- **COST ANALYSIS:** A cost analysis may be done comparing actual cost before and after implementation of the critical pathway. The financial department can often supply data or insight into cost issues that can be of great value in writing up a quarterly report.

FIGURE 4.4   *(continued).*

ance analysis, it is often the cost data that ultimately attract the most attention. Figure 4.5 includes some vital financial questions that should be considered when evaluating the progress of a critical pathway.

Once data are collected and analyzed regarding compliance with the critical pathway, achievement of patient outcomes, length of stay, patient satisfaction, and financial concerns, they can be compiled into report form and distributed to all involved with the critical pathway. The format of the report will change over time. In the beginning stages of implementation, a large portion of the report will be related to compliance with the critical pathway and meeting patient outcomes. This data will be used to improve the critical pathway and to change practice. As compliance becomes less of an issue, and more data are available regarding long-term patient outcomes, the report will outline issues that are related to this area.

The end result of outcome monitoring and variance analysis is the creation of an action plan to deal with the results. The action plan should always be documented and include specifics of what will be done to correct variances or improve patient outcomes.

CONCLUSIONS

Timely recognition of variances and addressing these deviations using a CQI format is the most important step that can be taken to assure success of critical path-

EVALUATING COST
◆ What is the cost of providing this service?
◆ How does the cost of providing this service today compare with the cost before implementing critical pathways?
◆ What are the five most significant charges under this DRG? How might those costs be limited?
◆ What are the most costly complications associated with this DRG?
◆ How often do these complications occur?
◆ What measures can be taken to prevent complications?

FIGURE 4.5 Evaluating cost.

ways and delivery of quality patient care. The critical pathway must be used to collect data about achievement of outcomes and patient response to treatment. Data need to be reviewed on a regular basis to determine if outcomes are consistently being met. Without some type of review process to address these issues, critical pathways are reduced to weightless practice guidelines. Consistent data collection and review of information on positive and negative variances will assure that both system and clinical advances are realized that will raise the level of practice and will benefit both current and future clients.

REFERENCES

Adams, C. E., & Biggerstaff, N. (1995). Reduced resource utilization through standardized outcome-focused care plans. *Journal of Nursing Administration, 25*(10), 43–50.

Berwick, D. M. (1988). Measuring health care quality. *Pediatric Review, 10,*11.

Crummer, M. B., & Carter, V. (1993). Critical pathways—the pivotal tool. *Journal of Cardiovascular Nursing, 7*(4), 30–37.

Docherty, J., & Dewan, N. (1995, October). *Guide to outcomes management.* Paper presented at the meeting of the National Association of Psychiatric Health Systems, Greater Bridgeport Mental Health Center, Bridgeport, CT.

Dziuban, S., McIlduff, J., Miller, S., & Dal Col, R. (1994). How a New York cardiac surgery program uses outcomes data. *Annuals of Thoracic Surgery, 58,* 1871–1876.

Fein, I. A. (1995). Information: Essential for survival. *Critical Care Management Consultants, 6*(1), 5–12.

Hronek, C. (1995). Redesigning documentation: Clinical pathways, flowsheets, and variance notes. *MedSurg Nursing, 4*(2), 157–159.

Krivenko, C., & Chodroff, C. (1994). The analysis of clinical outcomes: Getting started in benchmarking. *Joint Commission Journal on Quality Improvement, 20*(5), 260–266.

Larrabee, J. (1994). Using research to improve quality. *Nursing Quality Connection, 4*(3), 5.

Lumsdon, K., & Hagland, M. (1993). Mapping care. *Hospital and Health Networks, 67*(10), 34–40.

Mosel, D., & Gift, B. (1994). Collaborative benchmarking in health care. *Joint Commission Journal on Quality Improvement, 20*(5), 239–249.

Nadzam, D. (1994). The indicator measurement system: Change for the better. *Nursing Quality Connection, 4*(3), 6.

Patterson, P. (1994). Kaiser benchmarking study to identify best practice, *OR Manager, 10*(7), 16.

Schriefer, J. (1995). Managing critical pathway variances. *Quality Management in Health Care, 3*(2), 30–42.

Tallon, R. (1995). Devising and delivering objectives for disease state management. *Nursing Management, 26*(12), 22–24.

Zimmerman, J. E., Shortell, S. M., & Rousseau, U. M. (1993). Improving intensive care: Observations based on organizational case studies in nine units: A prospective multicenter study. *Critical Care Medicine, 21,* 1443–1451.

BIBLIOGRAPHY

Alba, T., Souders, J., & McGhee, G. (1994). How hospitals can use internal benchmark data to create effective managed care arrangements. *Top Health Care Finance, 21*(1), 51–64.

Buenos, M., & Hawing, R. Understanding variances in hospital stay. Nursing Management, *24*(11), 51–57.

Camp, R., & Tweet, A. (1994). Benchmarking applied to health care. *Joint Commission Journal on Quality Improvement, 20*(5), 229–238.

Campbell, A. B. (1994). Benchmarking: A performance intervention tool. *Joint Commission Journal on Quality Improvement, 20*(5), 225–228.

Coffey, R., Richard, J., Remmert, C., LeRoy, S., Schoville, R., & Baldwin, P. (1992). An introduction to critical paths. *Quality Management in Health Care, 1*(1), 45–54.

Denied, R. (1996). Data capture for quality management nursing opportunity. *Computers in Nursing, 14*(1), 39–44.

Dijerome, L. (1992). The nursing case management computerized system: Meeting the challenge of health care delivery through technology. *Computers in Nursing, 10*(6), 250–257.

Gottlieb, L., Sokol, H., Oates, K., & Schoenbaum, S. (1992). Algorithm-based clinical quality improvement. *HMO Practice, 6*(1), 5–12.

Hoyer, R. (1995). Prospective payment for home care. *Caring, 14*(3), 28–35.

Julian, K., & Poorer, C. (1991). Nursing case management: Critical pathways to desirable outcomes. *Nursing Management, 22*(3), 52–55.

King, M., McDonald, B., & Good, D. (1995). Redesigning care using total quality management and outcome/variance analysis. *Aspen's Advisor for Nurse Executives, 10*(5), 3–6.

Lenz, S. (1994). Benchmarking: Finding ways to improve. *Joint Commission Journal on Quality Improvement, 20*(5), 250–259.

Mendenhall, S., & Prock, S. (1995). Variance program melds hospital's, vendor's ideas. *Hospital Case Management, 3*(1), 1–4.

Norman, L. (1995). Computer-assisted quality improvement in an ambulatory care-setting: A follow-up report. *Joint Commission Journal on Quality Improvement, 21*(3), 116–130.

Nugent, W., & Schults, W. (1994). Playing by the numbers: How collecting outcomes data changed my life. *Annuals of Thoracic Surgery, 58,* 1866–1870.

Saul, L. (1995). Developing critical pathways: A practical guide. *Heartbeat, 5*(3), 1–10.

Shikiar, M., & Warner, P. (1994). Selecting financial indices to measure critical path outcomes. *Nursing Management, 25*(9), 58–60.

Simkin, B. (1995). Transitional pathway encompasses outpatient settings. *Hospital Case Management,* (Suppl. 1), 1, 12.

Tidwell, S. (1993). A graphic tool for tracking variance and comorbidities in cardiac surgery case management. *Progress in Cardiovascular Nursing, 8*(2), 6–19.

Windle, P., & Houston, S. (1995). COMIT improving patient outcomes. *Nursing Management, 26*(8), 64AA–64II.

Woodyard, L., & Sheetz, J. (1993). Critical pathway patient outcomes: The missing standard. *Journal of Nursing Care Quality, 8*(1), 51–57.

Zander, K. (1988). Nursing case management: Strategic management of cost and quality outcomes. *Journal of Nursing Administration, 18*(5), 23–30.

5

Critical Pathways in the Acute Care Setting

Barbara J. Lesperance

Demands of managed care and pressures from capitation have required hospitals to look closely at the way patient groups are treated in the hospital setting. Efficient, cost-effective, timely, coordinated care by all disciplines involved in the care of the patient is now required. The critical pathway serves as an ideal tool to map the sequence. While it may seem convenient to purchase pathways developed in other institutions, it is vital for each institution to design pathways appropriate to its unique patient population (Gordon, 1995; Reinhart, 1995).

This chapter is designed to assist novices as well as seasoned developers in creating effective clinical pathways for the acute care setting. The chapter will outline the sequence of work to be done, and by whom. Difficulties encountered will be reviewed, with suggestions on how to avoid or overcome them. Specific tools will be presented to assist planners in the implementation of critical pathways in the acute care setting.

DOING THE GROUNDWORK

Because the implementation of critical pathways requires a tremendous amount of commitment on the part of administration and staff on every level, it is essential that the proper groundwork be done before implementation begins. Done properly, this groundwork includes the following:

- establishing administrative commitment
- identifying areas for pathway development
- selecting key players

- identifying and defining benchmarks
- defining optimal practice on a time line
- removing barriers
- developing tools and forms

Thorough consideration of these areas will promote success of the critical pathway once implemented.

Establishing Administrative Commitment

The decision to develop critical pathways in a hospital setting is usually administrative. This resolution is often driven by the need to provide patient treatment in a time frame set by managed care companies, while demonstrating appropriate, fiscally responsible use of resources (Reinhart, 1995). A very real threat of business being taken away from certain physician groups or institutions demands lower cost while providing better service with superior outcomes (Korpiel, 1995).

The overall purposes for developing a pathway include reducing variability in care, identifying and conserving resource use, defining practice patterns related to outcomes, or defining practice patterns that can demonstrate cost/benefit performance for managed care companies.

The institution must be fully committed to support the use of pathways, with a physician skilled in quality improvement techniques appointed as chair for pathway development. Schriger (1994) suggests "the ideal candidate will have formal training in clinical epidemiology and/or statistics, familiarity with the medical field of the guideline, and experience in project management" (p. 116). Physicians are accustomed to autonomy; Wise and Billi (1995) noted that "a nonrandom survey of 290 physicians demonstrated that practicing physicians are generally suspicious of guidelines reported in the literature and are much more amenable to receiving new information after a local consensus is reached" (p. 467). It is especially important to have a physician who is respected in the organization taking a strong, visible leadership role in establishing practice guidelines and incorporating these guidelines into the critical pathway. The initial group should include a physician as leader, a clinical nurse administrator, and a system finance representative.

Identifying Areas for Pathway Development

The initial pathway should be written for an uncomplicated diagnosis. In the acute care setting, surgical procedures lend themselves to more structured treatment plans than are found with medical diagnoses that are often accompanied by complex comorbidities. Selection of the practice area, stated by a diagnosis related grouping (DRG), is guided by finance or quality improvement data, which include high volume, high cost, or variability in treatment. Symptoms such as pain can be selected, but require further definition (i.e., acute? chronic?) and may be more difficult to outline.

The initial pathway development group establishes agreement on what categories the critical pathway will address. Most pathways list days horizontally and care elements (consultations, treatments, tests, etc.) vertically. The goals of using critical pathways should also be established by this group. Reducing length of stay should not be an end in itself, although this may be realized once resource use is optimized. Goals for each path should reflect desired patient outcomes.

Selecting Key Players

The initial group should identify key players within each discipline who impact patient progress (Gordon, 1995; Reinhart, 1995). In a fractured hip pathway, emergency medicine, orthopedic surgeons, anesthesia, physical medicine, occupational therapy, orthopedic nursing personnel, and rehabilitation medicine all had significant roles. For the pneumonia pathway, emergency medicine, pulmonary and infectious disease specialists, internists, family practice physicians, and medical nursing units should be consulted. For chronic obstructive pulmonary disease (COPD), pulmonary medicine, family practice physicians, respiratory therapists, dieticians, medical nurses, physical therapists, and outpatient/inpatient pulmonary rehabilitation therapists are needed. Note that key representatives of each physician group who admit and treat that category of patient are included. It is vital to include representatives from major groups who will be expected to conform to guidelines if success is to be achieved.

The clinical nurse specialists (CNS) should also be included for each appropriate diagnosis. The CNS provides a unique *practically based* perspective that brings reality checks to the process. For example, in the development of a heart failure critical path, no mention was made of the standard of practice of using indwelling urinary catheters. The cardiologists and staff nurses accepted this as a given, while the infectious disease physician assumed no catheters would be used unless specifically required. The CNS heightened awareness that catheters were routinely used in the institution for this group of patients. By identifying this practice and questioning the need for catheters, the CNS was able to avoid the problems associated with implementation of a pathway that did not reflect current practice. In addition, a forum was provided where a practice pattern (the necessity of catheters) could be highlighted and earlier removal of catheters was promoted.

Identifying Desired Benchmarks

This phase of critical pathway development, the identification of benchmarks, is usually over a 2–3 month period. However, before identifying desired benchmarks, it is necessary to establish what is known about your institution's practice from financial services. Helpful information includes:

- numbers of admissions per month
- cost variability
- typical age range
- length of stay (is it significant by unit?)
- significant recidivism

Next, identify state and national figures for comparison. Obtain information from institutions or professional organizations known to hold the "Gold Standard" for care and compare elements of their processes and practices that positively impact desired outcomes. For example, in designing the hip replacement critical pathway at one local community hospital in Connecticut, practices at the Hospital for Special Surgery (New York City) were explored. In developing the path for pneumonia, newly published studies in the peer-reviewed *Journal of Pulmonary Medicine* were used to substantiate reduction of chest X rays and the need for early administration of intravenous antibiotics.

In addition, it is beneficial to recruit the assistance of medical records' personnel. It can be very helpful to identify which subdiagnoses (listed as ICD.9 codes) are included within a DRG and which criteria are used by coders to differentiate one DRG category from another. Current allotted length of stay for each diagnosis is helpful to know. For example, in researching DRG 14, cerebral vascular accident, it was found that the inclusion of the words "with infarction" in the medical record allowed double the revenue than if CVA alone were noted *assuming* an infarction.

A review of the literature should be accomplished before a comprehensive group convenes. Establishing research-based outcomes resulting from specific interventions is very beneficial when creating a critical pathway. Schriger (1994) recommends differentiating between intermediate outcomes (such as improved laboratory studies) and outcomes demonstrating clinical effectiveness of treatments (ability to perform ADLs (Activities of Daily Living or work). Literature outlining current therapies, tests that guide therapy and influence outcome, and references to outmoded treatments need to be collected. This information should be openly discussed and, if needed, debated in the multidisciplinary setting so that all disciplines and individuals will accept and follow the established guidelines.

Although retrospective chart review is recommended widely in the literature (Coffey, Remmert, Leroy, Schoville, & Baldwin, 1992; Goodwin, D., 1992; Saul, 1995), inclusion at this point is debatable. Chart reviews do provide information regarding current practice patterns, but the danger lies in that these same reviews may lead to an interpretation of current institutional practice as *the* desirable standard or benchmark. When a small group of nurses in our institution decided to initiate a path for myocardial infarction, they described current practice instead of researching controlled studies to define optimum practice. It may be wise when using data collected from retrospective chart reviews to present findings along with research-documented standards to avoid confusion.

In addition, printed guidelines may be collected from reputable sources to use as references. Well-researched guidelines are available for many diagnoses from the U. S. Department of Health and Human Services' Agency for Health Care Policy and Research, with references from an extensive literature review.

Defining Optimal Practice on a Time Line

When defining optimal practice, all identified key players should be convened. Care should be taken to enlist practitioners viewed as change agents within each group (Wise & Billi, 1995) and to avoid individuals known to be negative about any changes. The theme for this meeting should focus on patient outcomes and total quality initiatives to eliminate "turf" issues (Jones & Milliken, 1994). The information gathered from the preliminary group should be formally presented to introduce the "big picture."

Physician concerns of exposure to liability need to be addressed at this point (Korpiel, 1995). Statements disclaiming representation of paths as a standard of care but subject to judgments of skilled practitioners can be added to each path developed. Our institution opted to use the statement "Critical paths do not represent a Standard of Care. They are guidelines for consideration which may be modified according to the individual patient's needs." Since use of paths are relatively new, little precedent has been set in court cases. However, demonstrating the use of a researched team approach to care would seem to have a positive influence (Korpiel), particularly when deviations from the path are documented with supportive reasons (see chapter 9 for an in-depth look at liability issues).

The group should be asked if they feel other disciplines should be included in the critical pathway development. Sometimes case management links or discharge planners have important suggestions to add and should be contacted for input. Each discipline should be asked to meet in small groups to recommend their interventions in the desired time frame. In our hospital during the development of the pneumonia path, physical therapists felt their services would not be needed for the majority of clients. However, the group felt that a significant number of pneumonia patients may experience deconditioning and could benefit from therapy. The pathway thus suggests consideration of a physical therapy consult on day 3 if the patient is unable to tolerate activity progression. For heart failure, discussion centered around recommendations for cardiac rehabilitation versus physical therapy. The two groups discussed parameters for each specialty and made these parameters part of the decision tree in the path.

Team members must be committed to the development and subsequent implementation of the clinical path (Korpiel, 1995). Each representative group will be asked to research further (if needed) specific aspects of care related to their specialty and to return with recommendations to include in the path. The best time for the interventions and desired outcomes should be outlined. Groups should be cautioned to include only necessary services that provide significant health benefits (such as longevity or improved quality of life) and not interventions that have been customary but with no demonstrated impact on outcome (Hadorn, 1994). For example, in our hospital, drawing of hemoglobin and hematocrit levels had been routinely performed every 6 hours on patients admitted with gastrointestinal bleeding. This practice was demonstrated to be of no use in guiding therapy, was costly, and actually was more likely to reduce the patient's blood levels, albeit by small increments. Similarly, the routine ordering of guaiac testing of stools in patients known to have heme-positive stools was found to be wasteful. Any tests included in the path should be directly linked to the timing of treatment decisions (Eddy & Hasselblad, 1994).

Teams should be given the anticipated date for regrouping , usually 4 to 6 months (Gordon, 1995). The physician team will likely need the most time to come to a consensus on essential care components. A physician leader not directly involved in practice issues but skilled in quality improvement issues should serve as moderator for the physician group. It is helpful for a clinically astute member of the initial group (often the clinical nurse specialist) to be present at the work sessions to maintain the focus on achieving outcomes. When a consensus cannot be reached on a specific intervention, the 80-20 rule should be followed: listed interventions should apply to 80% of the population, with 20% expected to show some variance.

Removing Barriers

Each discipline should also consider the current hospital systems and ask themselves: "Will current systems allow interventions at the appropriate time, or are changes needed?" When our hospital recommended surgical intervention for fractured hips on day 2, policies needed to be changed to allow the procedure to be scheduled on weekends. When ambulating COPD patients with oxygen were added to the pathway, nursing units were provided with rolling oxygen tank holders. The inclusion of representatives from medical records may assist in the approving of forms to be added to the permanent medical record. As mentioned earlier, a major barrier to the success of a critical pathway is physician/nurse buy-in. Time spent eliminating negativism of practitioners and convincing them of the benefits of a critical pathway will yield huge benefits during implementation.

TOOLS AND FORMS

The following tools and forms should be considered when developing a critical pathway for use in the acute care setting:

- structured path
- standard order sheet
- patient critical pathway

- teaching materials
- benchmarking tools
- follow-up letters

Working out in advance which tools and forms will accompany the critical pathway can save time and prevent problems during the implementation phase.

Structured Path

When the multidisciplinary group reconvenes, they must review the structure of the path and agree on time frames for interventions, aiming toward efficient patient management and multidisciplinary integration. Sometimes the group will recognize the need for a decision tree or algorithm to direct the appropriateness of specific interventions. This need can be met through a practice guideline (Wise & Billi, 1995).

Some of the critical paths in our institution have combined the critical pathway with the decision tree of a practice guideline. For example, for treating patients with heart failure, an algorithm assisting in treatment decisions based on measures of dyspnea and resolution of symptoms was added to the pathway (Figure 5.1). Cardiologists and internists also came to an agreement and listed indications for cardiology consultations for patients with heart failure. With the pneumonia critical pathway, infectious disease specialists added a guideline that listed specific antibiotics for initial therapy dependent on community versus institutionally acquired pneumonia. Gastroenterologists listed specific parameters in a guideline to determine appropriateness of ICU (Intensive Care Unit) versus floor treatment.

Standard Order Sheet

The creation of a standard order sheet (Figure 5.2) incorporating blood tests, consultations, medications, activities, and treatments can assist in directing the timeliness of critical pathway recommendations. Standard orders can be particularly helpful in quality improvement initiatives. In our institution, patients with deep vein thrombosis were noted to have prolonged stays due to delays in achieving adequate oral anticoagulation. This was directly related to the delay in administering oral anticoagulants until the 2nd or 3rd day of hospitalization. Preprinted standardized orders included a weight-based bolus of heparin, with initiation of oral anticoagulation on the day of admission. These strategies resulted in desired levels of anticoagulation in 87% of the patients within 24 hours of therapy (Kotch, 1995).

Patient Critical Pathway

Critical pathways written specifically for the *patient* follow the time line incorporated into the critical pathway. Our institution removed the word "critical" from the patient document to avoid misrepresentation of severity of illness. The patient path outlines

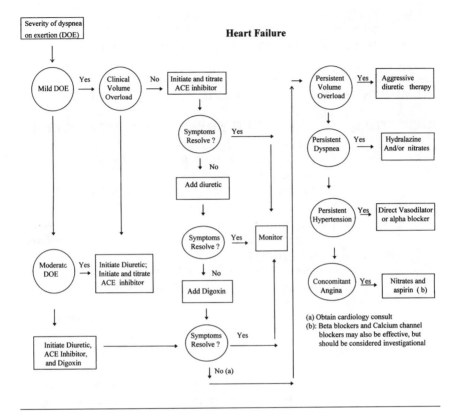

FIGURE 5.1 Algorithm for heart failure.

Note: Used with permission of Danbury Hospital, Danbury, CT.

expected daily events, mirroring elements on the critical pathway, such as assessments by therapists, expected changes from intravenous to oral medications, and diet teaching (Figure 5.3). The patient path is given to the patient on admission. If the patient is too ill or otherwise unable to use the patient path, the material is given to the significant others.

The purposes of the patient path are twofold: to inform the patient about his or her hospital stay and to allow and encourage the patient to become a partner in his or her care. If on day 3 the path outlines changing an intravenous antibiotic to an oral form, the patient who is still receiving the intravenous form would be encouraged to ask the physician to explain the reason. Similarly, if a listed visit from an outpatient pulmonary rehabilitation nurse is not realized, the patient's request can reinforce the need for the consultation.

Patient pathways should be written in appropriate language and print. Consideration should be given to providing paths in languages to meet the institution's client

Use Ball Point Pen, Bear Down Firmly
(CROSS OUT ANY ITEMS PREVIOUSLY PERFORMED OR
WHICH YOU DO NOT WISH TO ORDER & INITIAL)

STANDARD ORDERS: **VENOUS THROMBOEMBOLISM/
PULMONARY EMBOLISM STANDARD TREATMENT ORDERS**

DATE   TIME   1. Baseline H & H, platelet count, INR, APTT, UA, rectal or
                    guaiac stool
                    Patient weight **in kg:** Actual ___   Estimated ___
                    Desired APTT range : 50–80 sec
                 2. *Bolus dose: _____Units heparin, IV push
                    After Baseline Labs Are Drawn
                 3. Begin IV: 20,000 Units heparin in 500 cc D5W (40
                    units/cc), starting at **33ml/hr**(1,320 U/h), **immediately**
                    after giving bolus.
                 4. Obtain APTT **STAT** 6h after bolus dose.
                 5. Adjust drip based on the following nomogram (sliding
                    scale): **

APTT	OTHER	RATE CHANGE: ml/hr
<40	**call MD,** repeat initial bolus	increase by 6 cc/hr
41–49		increase by 3 cc/hr
50–80		**no change**
81–90		decrease by 3 cc/hr
91–105	**hold 60 min; restart**	decrease by 3 cc/hr
>105	**hold drip; call MD *****	

                 6. Obtain APTT **STAT** 6h after each dosage change, adjust-
                    ing heparin drip according to the sliding scale until APTT
                    is therapeutic (50-80 sec). **Draw one additional APTT
                    6h following the <u>first</u> therapeutic result.**
                 7. **Obtain the following labs: Daily INR and APTT; qod
                    H & H with platelet count; daily heme test urine and
                    emesis; guaiac stools at least once daily.**
                 8. **All rate changes must be written on the Doctor's order
                    sheet at the time of each change.**
                 9. Coumadin ___ mg at 8pm on **day 1.** Ë (Maintenance
                    dose INR target = 2.0–3.0)
                    Coumadin ___ mg at 8pm on **day 2.**
                10. Bedrest for 3–5 days, elevate knees and feet to 45 degrees.

         M.D. Signature/ID#: _____

FIGURE 5.2   Doctor's orders.

STANDARD ORDERS *(continued)*.
    M.D. Notes:

          *   Recommended bolus: 80 units/kg body weight (Ann Intern Med., 1993;119:874–881)

          **   Suggest discontinuing heparin after 5 days.

          ***Consider: Repeat **STAT** PTT in 1 hour. If value <105, continue standard nomogram. If value >105, hold 1 more hour and repeat **STAT** PTT until value is <105.

          Ë   Concomitant anticoagulation with Coumadin should be initiated on day 1 of heparin therapy. In massive PE, Coumadin is delayed until day 3–5.

FIGURE 5.2   *(continued)*.

Authorization is hereby given to dispense medication in accordance with the Hospital Formulary System. 1/96. bl:A:\DOCORD[DH# 85877].

Note: Used with permission of Danbury Hospital, Danbury, CT.

needs. Stewart (1996) notes that "54% of American adults can't read or are marginally literate" (p. 32j). Stewart recommends using computer software that measures readability, or using the SMOG formula (a formula with a conversion table that predicts readability and grade level of reading material) to predict difficulty levels of written material. Consideration should be given to the age of clients. Our patient paths use 14 point size print to maximize ease of reading for the older adults.

Teaching Materials

Teaching protocols are standard in our institution, outlining content to review with patients. Figure 5.4 shows the teaching protocol for heart failure. The initial page assessing readiness to learn also meets requirements of state and federal accrediting agencies. These not only assist the nurses in documenting teaching but also serve to familiarize nurses with diagnostic-specific information. Included are food and drug interactions, symptom recognition, and activity recommendations in addition to descriptions of the disease process. Additional handouts provided by drug companies or associations such as the American Lung or the American Cancer Society are provided when appropriate.

Benchmarking Tools

Tracking appropriate use of the recommended critical pathway elements can be done in multiple ways. Ideally, initialing completion of an intervention on the path itself is most desirable, with any deviation (variance) from the critical pathway

Heart Failure

FROM HOSPITAL TO HOME;
YOUR DAILY PLAN OF CARE

Patient Path

	DAY 1
CONSULTATIONS:	Your nurse will ask you about your past health and current illness. It is important to tell your nurse about any recent changes in your sleep patterns, medicines, or diet. Your doctor will examine you and may have a heart specialist come and see you.
TESTS:	Blood tests will be drawn to check for any damage to your heart and to check for electrolytes (substances in your blood that help your body function correctly). These may be repeated later today.
ACTIVITY:	You may need extra rest today. Ask your nurse if you are permitted to use the bathroom. Ask your nurse about any activities you should not do. Call your nurse to assist you out of bed. Some people feel dizzy or light-headed from the medicines given or from the fluid loss. To avoid this, please get up slowly and sit at the edge of the bed for a few minutes before standing.
TREATMENTS:	Your pulse and blood pressure will be checked every 4 hours (or so). You may receive oxygen through the nose. You will be weighed today and every morning. The nurse or assistant will use the same scale to check your weight. You may have an electrocardiogram (EKG) done to check your heart. You may be placed on "telemetry"—it allows your nurse and doctor to monitor your heart's rate and regularity.
MEDICATIONS:	You will receive medicines by I.V. today, especially ones to reduce the amount of excess fluid in your body. Your pulse and blood pressure will be checked. You will receive any other medications prescribed by your doctor.
DIET:	You will be given a diet low in salt. Your nurse will explain any other special diet needs to you today.

FIGURE 5.3   Patient pathway.

DISCHARGE PLAN:	A case manager may contact you to help plan your discharge from the hospital.
TEACHING:	It is important to tell your nurse right away if you feel any chest pressure or pain.
	DAY 2
CONSULTATIONS:	
TESTS:	You will be weighed today to check how much fluid you have lost. Tests may be ordered by your doctor. Your nurse will give you information about any tests. Blood tests will be done once or twice today.
ACTIVITY:	Today you can get up and walk around your room if you feel able. Be sure to rest whenever you start to feel tired. Ask for assistance before getting up and get up slowly.
TREATMENTS:	Pulse and blood pressure will be checked about every 8 hours (each shift). You may still be getting oxygen through your nose.
MEDICATIONS:	You may be switched to medicines by mouth today.
DIET:	You will continue on your low salt diet. The dietician may see you to discuss foods to avoid to help prevent problems with holding onto fluid.
DISCHARGE PLAN:	The case manager will probably contact you to discuss needs you may have after your discharge. It is important to discuss how you will manage meals, bathing, getting in and out of bed, and getting to the bathroom after you leave the hospital.
TEACHING:	The nurse will show you how to pace your activities to avoid fatigue. Your nurse will discuss your medicines and the timing of your medicines, such as before or after meals. Your nurse will begin to discuss signs and symptoms of CHF. You should report any difficulty breathing or any chest pressure to your nurse and/or doctor. Your nurse will ask you about any medicines you take that are not ordered by your doctor: medicines for upset stomach, headache, or muscle aches.
	DAY 3
CONSULTATIONS:	No consultations are scheduled for today.
TESTS:	You will be weighed today. Look at the chart at the back of this book. Write down your weight on this chart and con-

FIGURE 5.3   *(continued).*

	tinue to use this chart at home. Your weight may be different on your home scale. Discuss this with your nurse. Blood tests may be done.
ACTIVITY:	Your nurse will help you walk in the hall 10-20 feet twice. Let your nurse know if you become tired.
TREATMENTS:	The nursing staff will continue to check your blood pressure today.
MEDICATIONS:	Your medicines may be changed to pills today, if they weren't changed yesterday. Your nurse will give you information about your specific medicines today.
DIET:	Your diet will be reviewed by your dietician and nurse. You will be told about any foods to avoid.
DISCHARGE PLAN:	Review with your nurse and case manager any questions or problems you may have concerning your home situation. Your nurse will review your discharge plans with you. You may be discharged today.
TEACHING:	Your nurse will review your medicines and any drug interactions. Your nurse will give you a pamphlet about CHF. Please ask your nurse if you do not understand any information. Be sure to weigh yourself **at the same time every day** using **the same scale.** It is important to call your doctor if you gain **2 pounds in 1 day, or 3 to 5 pounds in 1 week.** You should also call your doctor if you need more pillows to sleep comfortably or if you are waking up at night unable to breathe. Let your doctor know if you see water buildup in your ankles or legs.
	DAYS 4–7
CONSULTATIONS:	No consultations scheduled for today.
TESTS:	No tests scheduled for today.
ACTIVITY:	Continue to walk with your nurse in the hallway, 10-20 feet, two to three times.
TREATMENTS:	Pulse and blood pressure will be checked two or three times today.
MEDICATIONS:	Continue your oral medicines.
DIET:	Continue your diet.
DISCHARGE PLAN:	Your doctor may discharge you today. Be sure your nurse reviews your discharge plan.
TEACHING:	Your nurse will review your medicines and their side effects. Your nurse will review again the signs and symptoms of CHF, and when you should call your doctor.

FIGURE 5.3   *(continued).*

STAYING WELL:	You should be breathing better now and feeling well. CHF is a condition that cannot be "cured," but can be controlled. To prevent recurrence, you must take your medicines every day as prescribed. Exercise can help you get stronger. Be sure to ask your doctor about any activity/exercise program. Discuss with your doctor getting a flu shot or pneumonia vaccine. Rest periods should be planned in your day. Discuss this with your nurse or doctor.

FIGURE 5.3 *(continued).*

Note: Used with permission of Danbury Hospital, Danbury, CT.

noted in the progress record (Hronek, 1995). This method requires that the critical pathway be a permanent part of the official medical record. The major advantage is providing ease of documentation. Gordon (1995) states, "If the nurses will be doing the documentation, the forms must be 'in place of' a charting form rather than 'in addition to'. Double and triple charting requirements drive nurses crazy" (p. 199). When the path is not an official part of the record, an alternative record of monitoring is advisable. Our institution is still using separate monitor sheets, outlining achievement of desired outcomes. We are currently piloting a critical pathway for heart failure that incorporates charting and noting of variances directly on the critical pathway (Figure 5.5). If the pilot proves successful, all current critical paths will be rewritten to conform to this model.

Since the critical pathways are interdisciplinary, it is appropriate for each discipline to note compliance with pathway recommendations. In analyzing the pneumonia critical pathway, our institution wanted to verify administration of the intravenous antibiotic within 4 hours of patient arrival. Initial analysis noted a discrepancy in the time of antibiotic administration; this uncovered the need to identify whether the patient was a direct admit or was admitted through the emergency department. The process responsible for the variance was then systematically addressed.

At the time of discharge, the completed monitor sheet is placed in a designated notebook and is analyzed monthly. Figure 5.6 is the monitor used for the pneumonia critical pathway, identifying outcomes our institution wishes to track. Fully computerized systems can incorporate this information in a database.

Follow-up Letter

To assess understanding of medication schedules, diet, and the need for daily weights, our institution is piloting a letter given at discharge to patients hospitalized with heart failure (Figure 5.7). The patient's follow-up visit is noted on the letter, with instructions to return the letter at the first visit.

A simple weight chart is included in the patient path (Figure 5.8).

Topic	Methodology/Content	Patient/S.O. Response & Comments	Patient Outcomes	Date/Initials Started	Date/Initials Completed
Readiness to learn	Review: Patient readiness to learn		( ) Pt. ( ) S.O. States/exhibits readiness to learn by: 1. Ability to concentrate. 2. Willingness to learn. 3. Ability to communicate.		
Disease: Upper and Lower GI Bleed	**Discuss disease process and causes for:** UPPER GI BLEED • Esophagitis( ) • Gastritis( ) • Mallory-Weiss Tear( ) • Peptic Ulcer( ) • Stress Ulcer( ) LOWER GI BLEED • Anal fissures( ) • Angiodysplasias of colon( ) • Cancer of colon and rectum( ) • Diverticula of colon( ) • Hemorrhoids( ) • Polyps( ) • Rectal trauma and proctitis( ) • Ulcerative colitis( ) • Rectal fistulas		( ) Pt. ( ) S.O. Verbalizes understanding of disease process and probable cause		
Tests	**Discuss:** • Endoscopy( ) Handout( ) Upper( ) Lower( ) • Bleeding Scan( ) Usually done when actively bleeding: May need more films in 24–48 hours.		( ) Pt. ( ) S.O. Identifies procedural steps for _____ _____ _____		

FIGURE 5.4 Patient teaching protocol.

Topic	Methodology/Content	Patient/S.O. Response & Comments	Patient Outcomes	Date/Initials Started	Date/Initials Complete
	• Angiogram( ) Handout( ) • Other				
Diet	**Discuss:** • Diet plan • Gastric and/or esophageal irritants • Alcohol • Medications such as Aspirin or NSAIDS • Personal triggers • Smoking		( ) Pt. ( ) S.O. Verbalizes diet plan ( ) Pt. ( ) S.O. Identifies specific substances to avoid ( ) Pt. ( ) S.O. Verbalizes plan to change lifestyle		
Medication	**Discuss:** • Medication: _____ • Reason: _____ • Dose: _____ • Frequency: _____ • Food & drug interactions: _____ • Major side effects: _____ • Special instructions: _____		( )Pt. ( ) S.O. Identifies medication, side effects, and interactions for all medications prescribed.		

FIGURE 5.4 *(continued).*

*S.O. = Significant other.

Note: Used with permission of Danbury Hospital, Danbury, CT.

	DAY 1/Date:	DAY 2/Date:	DAY 3/Date:	DAY 4/Date:
	Yes No □ □ List problems not related to heart failure on care plan/ flow sheet. □ □ Obtain list of all meds.			
TESTS	Initials: Yes No □ □ Blood wok complete: □ □ Electrolytes normal: □ □ Cardiac Enzymes normal:	Yes No □ □ Lytes normal □ □ Echo EF____		Yes No □ □ Lytes normal
ACTIVITY	Yes No □ □ OOB as tolerated, BRP	□ □ Ambulates to ___ ft. If unable to ↑ activ, consult PT. □ □ PT notified	□ □ Ambulates to ___ ft.	□ □ Ambulates to ___ ft. □ □ Cardiac rehab candidate
TREATMENTS	Yes No □ □ Weight: ___ Scale #: ___ Time: ___ □ □ Pulse ox: ___ Time: ___ □ □ Nasal O₂ if pulse ox <92 L/m Urine Output/Diuresis: Days ___ Eve ___ Nite ___	Weight: ___ Scale #: ___ Time: ___	Weight: ___ Scale #: ___ Time: ___	Weight: ___ Scale #: ___ Time: ___
MEDICATIONS	IV kiuresis until volume overload resolves. Ace inhibitors (short acting) Digoxin (in severe CHF)	□ □ Change to po diuretic.	List Med Changes	

FIGURE 5.5 (FRONT). Nursing documentation for heart failure critical pathway: DRG #127.

DIET	2 Gm Sodium:___	2 Gm Sodium:___	☐ Diet Consult	2 Gm Sodium:___
DISCHARGE PLAN	Yes No ☐ ☐ Evaluate information from data base	Yes No NA ☐ ☐ ☐ Home Assist Identified	☐ ☐ Case Management Date:___	☐ ☐ Home plan established
TEACHING	Yes No ☐ ☐ Pt. instructed to call nurse for chest pain, SOB	Yes No ☐ ☐ Review S/S ☐ ☐ Review OTC meds ☐ ☐ Review activity pacing/avoid fatigue	☐ Specific med teaching ☐ Review S/S heart failure	☐ Review/complete teaching protocol.
OUTCOMES	Yes No ☐ Patient Path given and reviewed ☐ No evidence of chest pain ☐ Evidence of diuresis ☐ Pulse ox >92%	☐ Weight loss ☐ OOB,BRP ☐ EF known ☐ Lytes WNL	☐ Weight loss ☐ Ambulates to 50' ☐ Verbalizes s/s HF.	☐ ☐ States reportable s/s ☐ ☐ Verbalizes approp. diet choices. ☐ States sched/pupose of meds. ☐ States plan for daily wt
NURSE INITIALS:	DAY___ EVE___ NITE___	DAY___ EVE___ NT___	DAY___ EVE___ NT___	DAY___ EVE___ NT___

FIGURE 5.5 (FRONT). *(continued).*

Variance Notes

Codes:	(P) = Patient Related		(S) = System Related		(C) = Caregiver Related	
Time & Code	Date _____	Time & Code	Date _____	Time & Code	Date _____	Time & Code

(FIGURE 5.5 BACK) Nursing documentation for heart failure critical pathway: DRG#127.

Note: Used with permission of Danbury Hospital, Danbury, CT.

PNEUMONIA: DRG 89

Patient on Critical Path: Yes No	
QI Completed by Staff Nurse: Yes No	
Admit Date:_____ **Day of Week:** (circle) M T W T F S S	
Arrival Time In ED:_____ **Time to Floor:** _____ **Direct Admit:** Yes No	
Admitted From: Home Group Home ECF Other: (specify) _____	
Disposition: Home Group Home ECF Other: (specify) _____	
Time IV antibiotic ordered:	
Time was given (circle ED or floor)	ED _____am/pm Floor_____am/pm
Day IV antibiotic changed to PO	(Circle) Day: 1 2 3 4 5 >5
Documented Need for continuance of IV antibiotic?	Yes No
ID/Pulm consult Day 3 if condition worsening	Yes No
Total # CXR: List <u>date</u> of each CXR	
Circle day 02 off (Circle E if none used)	(Circle) Day: 1 2 3 4 5 >5 E
Changed to Saline Lock?	(Circle) Day: 1 2 3 4 5 >5 E
Sputum Sample Sent From:	ED Floor None
Respiratory Therapy Consulted for Sputum?	Yes No
Pneumonia Teaching Protocol completed?	Yes No
If protocol not complete, is reason documented?	Yes No
Documentation of patient response?	Date:____ Comment:_____
Patient verbalizes need for rest, fluid, completing antibiotics?	Yes No
Patient functionally independent at d/c?	Yes No (If no, what arrangements made?)
Any problem delaying d/c?	Yes No (If yes, explain)

FIGURE 5.6 Patient care monitor.

Note: Used with permission of Danbury Hospital, Danbury, CT.

It is hoped that misunderstanding of the treatment regimen can be discovered early and clarified to prevent readmission of patients due to lack of adherence to the prescribed therapy. Also at discharge, a letter is sent to attending physicians with information on desired treatment outcomes stressed during the hospitalization. The physicians are reminded to look for patient letters to assess treatment adherence.

The specific forms used for each critical pathway should be guided by the outcomes desired. For example, there are several forms described for use in conjunction with the heart failure critical pathway not used with other paths. In developing the heart failure critical pathway, consideration was given to the primary goal of

Dear _____,

You have been in the hospital because of heart failure. To stay well and to avoid a repeat hospital admission, it is very important to take the medicines listed on your discharge sheet from your doctor, **even if you are feeling better.**

You are to see Doctor _____ in the office on _____ at _____.
Please call _____ if you need to change the time or date.

1. Bring the **weight chart** with you to your visit, and the list of medicines which you take.

2. Write down the **medicines** you use during your first full day at home in the chart below:

Medicine	Number of pills at each dose	Times per day taken	Any problems or difficulties

3. Please be prepared to discuss your **typical daily food and beverage intake.**

   Notes/questions to ask:

   _____

4. **Activity:** How much I can do (walking, house work, other). Please write down any trouble breathing. How many pillows do you need to sleep? Do you need more now than you did before?

   _____
   _____
   _____

FIGURE 5.7   Patient follow-up letter.

Note: Used with permission of Danbury Hospital, Danbury, CT.

decreasing recidivism. Heart failure has been identified as "the most important public health problem in cardiovascular medicine" (Field, 1994, p. 2). Readmission of patients with heart failure ranged from 19% to 42% in a variety of hospitals reported in a 1994 Cardiology Preeminence Roundtable (Field). That roundtable reported an emerging "Gold Standard," listing an "opportunity to reduce inpatient CHF admissions by as much as 50%" (p. 10). Research shows that readmissions resulted from up to 41% of patients discontinuing or reducing prescribed dosages of medication (Field, 1994) and failure to recognize symptoms early

(Use the same scale at the same time each day)

SUNDAY	MONDAY	TUESDAY	WEDNESDAY	THURSDAY	FRIDAY	SATURDAY

FIGURE 5.8   Weight chart.

enough to allow timely outpatient help from caregivers. Thus our critical pathway focuses on methods to reinforce patient teaching and enhance adherence to and understanding of the medical plan of care.

IMPLEMENTING THE CRITICAL PATHWAY

A structured educational program to introduce the developed critical pathway and its forms to all disciplines is imperative (Reinhart, 1995). Departmental meetings can be used to present the critical pathway elements and goals to physicians. Physicians instrumental in critical pathway development should present the current status of patient treatment and stress the optimal recommended standards as they apply to the local situation (Wise & Billi, 1995). Wise and Billi reported survey results from eight organizations, including the American College of Physicians, the American College of Cardiology, and Harvard University, noting that these organizations were *not* using as many resources in *implementing* the guidelines as they used in *developing* the guidelines. Research findings that support elimination of unnecessary tests should be reviewed, as well as support for specific interventions included in the critical pathway. In a teaching institution, care should be taken to include house staff who will be expected to use the pathway in guiding patient care.

Nurses and unit secretaries must attend inservices describing the use of the critical pathway and its elements, plus outcome goals desired for the patients. A review of the disease process may be needed to validate the addition of some interventions. Surprisingly, inservices for the COPD critical pathway uncovered inconsistencies and lack of understanding by nurses in the correct use of metered dose inhalers (timing and method). The value of the critical pathway in structuring the patient's hospital stay, providing interventions in a timely manner, and assuring achievement of the maximum functional ability of the patient must be outlined.

Specific roles of nurses in documenting progress on the critical pathway must be described and demonstrated. Since the pathway was developed as a multidisciplinary effort, using multidisciplines to review the integration of their roles is helpful. Inservice presentations should be done within 1 to 2 weeks of initiating the path to maximize retention.

Posters "advertising" the start of a path may be effective reminders, as can messages that appear on the computer screen in the patient-care areas. A system of demonstrating where forms will be found, and where completed forms are placed, is needed. In our institution, all critical pathways and supportive forms were paper clipped together to maximize ease of path use. Secretaries were asked to place the critical pathway forms on the nurses' clipboard at the time of admission.

Support for the staff and continued praise and encouragement are critical to achieving success in implementing the critical pathway. Guidance in using forms correctly during initial implementation saves many hours of correcting habits later.

EVALUATING OUTCOMES AND FEEDBACK

Concurrent chart review assures that the practitioners are using the critical pathway to guide interventions during the patient stay and helps to integrate the pathways into practice patterns. For example, in our institution, initially, the house staff and attendings needed frequent reminders to limit chest films in pneumonia patients who showed clinical improvement. After about 9 months, the house staff progress notes responded to requests by the attendings to repeat chest films with notations such as "patient improving—repeat films would be of no benefit in therapy at this time."

If critical paths are new to your organization, cries of "cookbook medicine" can be anticipated. Physicians like to focus on the *art* of medicine, but over time, practice patterns by individual physicians clearly emerge. Critical paths are designed to *change* and optimize practice patterns. Paths should be revised as new research gives support to include or delete interventions.

Assessment of problem areas should be done by factual reporting to avoid finger-pointing or blame. No conclusive data collection should be done the first 2 or 3 weeks of critical pathway implementation. Encouragement, interpretation of the recommendations, and review of the desired outcomes are appropriate efforts the first few weeks. Some physicians or nurses will comment that an individual patient is "too sick" to be placed on a critical pathway. Reinforcement of the need for recognition of acuity and documenting the variances from the critical pathway can overcome this attitude.

After the initial few weeks, reports should be generated and distributed to all involved disciplines. Initial reports may show positive results from shorter length of stays or reduced use of services, but long-term outcome achievements may take a year or more to report. Initially, greater costs may be incurred (home teaching for heart failure patients, cardiac rehab exercise programs, outpatient smoke cessation programs), and the effect of these interventions on outcomes (less recidivism or reported improvement of quality of life) cannot be measured for a year or more.

Written factual reports provide a positive means of gaining support of providers who may not have been aware of their practice. The manager of our emergency department was certain that antibiotics were administered promptly for patients diagnosed with pneumonia in the emergency setting. When concurrent reviews demonstrated only 40% compliance with antibiotic administration in the emergency department, greater effort at meeting this standard was seen. Compliance improved to 60% over the following months, and a goal of 75% is in place for the next quarter.

At the final meetings of critical pathway completion, the decision should be made regarding who will track the variations and outcome measurements for the path. Without a commitment from the project team, the duty often falls on the nurses. In fully computerized systems, tracking can be integrated into the system with periodic evaluations to apprise the necessity of measuring certain parameters. In our institution, the COPD path required notation of administering intravenous corticosteroids on admission. There was 100% compliance with this measure over the first 6 months of the path. We therefore removed this from the monitor.

CONCLUSION

Acute care is now subject to more scrutiny than ever before. Under any capitation program, even *short* acute care stays will be a major drain on a facility. Critical pathways can provide a structured approach to maximize progress toward desirable outcomes while controlling resource use. Very simply, the critical pathway provides a road map for a designated trip into the health care system with the engine tuned to maximum efficiency. Critical pathways developed from a research base are an institution's best means of knowing *how* and *why* clients are managed during an illness episode.

Acute care settings can no longer function in isolation. Health patterns of clients prior to hospitalization must be addressed, and steps to maintain wellness must be initiated for the post-hospitalization phase. Elements of this "seamless care" can be added to the critical pathway and communicated to the continuing care agencies. Continuous communication regarding progress of outcome achievement must be presented frequently and in as positive a manner as possible. Not all diagnoses will have critical pathways written for them. This is neither advisable nor appropriate. Critical pathways work best when change is needed, either in outcome achievement or resource allocation. When developed and utilized in this framework, critical pathways have the potential for becoming the tool that will provide for "seamless transition" of clients from acute care to other settings, now and into the next century.

REFERENCES

Eddy, D., & Hasselblad, V. (1994). Analyzing indirect evidence. In K. McCormick, S. Moore, & R. Siegel (Eds.), *Methodology Perspectives* (pp. 5–13). Washington, DC: U.S. Department of Health and Human Services.

Field, J. (1994). Beyond four walls: Research summary for clinicians and administrators on CHF management. *Cardiology Preeminence Roundtable*. Washington, DC: Advisory Board.

Goodwin, D. (1992). Critical pathways in home care. *Journal of Nursing Administration, 22*(2), 35–40.

Gordon, M. (1995). Steps to pathway development. *Journal of Burn Care & Rehabilitation, 16*(2, Pt. 2), 197–202.

Hadorn, D. (1994). Use of algorithms in clinical guideline development. In K. McCormick, S. Moore, & R. Siegel (Eds.), *Methodology Perspectives* (pp. 93–104). Washington, DC: U.S. Department of Health and Human Services.

Hronek, C. (1995). Redesigning documentation: Clinical pathways, flowsheets, and variance notes. *MedSurg Nursing, 4*(2),157–159.

Jones, R., & Milliken, C. (1994). Collaborative care: Pathways to quality outcomes. *Journal of Healthcare Quality, 16*(4), 10–13.

Korpiel, M. (1995). Issues related to clinical pathways: Managed care, implementation, and liability. *Journal of Burn Care & Rehabilitation, 16*(2, Pt. 2), 191–195.

Kotch, A. (1995). Verbal communication. Danbury, CT.

Reinhart, S. (1995). Uncomplicated acute MI: A critical pathway. *Cardiovascular Nursing, 3*(1), 1–7.

Saul, L. (1995). Developing critical pathways: A practical guide. *Heartbeat, 5*(3), 1–12.

Schriger, D. (1994). Training panels in methodology. In K. McCormick, S. Moore, & R. Siegel (Eds.), *Methodology Perspectives*. Washington, DC: U.S. Department of Health and Human Services.

Stewart, K. (1996, January). Written patient-education materials: Are they on the level? *Nursing, 96,* 32–34.

Wise, C., & Billi, J. (1995). A model for practice guideline adaptation and implementation: Empowerment of the physician. *Joint Commission Journal on Quality Improvement, 21*(9), 465–476.

6

Critical Pathways in Ambulatory Care

Patricia C. Dykes

THE CHANGING HEALTH CARE ENVIRONMENT

All areas of health care are changing at a dramatic pace with the ambulatory care setting reflecting climactic growth since there is a shift of delivery of care from inpatient to outpatient settings. Between 1985 and 1992, the number of inpatient admissions to short-stay hospitals decreased by 8% to 32.6 million in 1992. During the same period, outpatient visits in short-stay hospitals grew by 50% to 409 million in 1992 (National Center for Health Statistics, 1992). In 1989, one-half of all surgery in short-stay hospitals was performed on an outpatient basis, 3 times the proportion of outpatient surgery in 1980 (National Center for Health Statistics, 1992). Ambulatory care currently represents 30% of hospital revenues, and is estimated to account for 50% of hospital revenues by the year 2000 (Howe, 1996).

Economic shifts as well as population shifts are taking place. The payor shift is away from private insurance toward managed care contracts and ambulatory patient groups (APGs). This change has necessitated a downward spiraling of costs in order for ambulatory service centers to compete effectively for business. In ambulatory care, as in all health care settings, the challenge is: "How do we continue to provide high quality care in an abbreviated period of time and with less resources than we had ever imagined possible?"

This question rings true particularly in the outpatient setting. Practitioners must provide care in a limited amount of time, and document their outcomes without the luxury of having the patient around to address variances. There is no "down" time in an ambulatory setting to allow staff to catch up on medications that may not have been given or treatments that may not have been done. By organizing care that is to be given into essential interventions and desired patient outcomes, critical pathways provide a means to transcend the pressured health care ambience to deliver distinctive, yet fiscally sound, health care.

As care continues to be delivered more often in outpatient settings and less often in acute care centers, the staff in ambulatory care centers are faced with the burden of learning how to care for increasing numbers of patients with diagnoses that were exclusive to acute care settings in the recent past. There are few resources to provide in-service education on proper care of patients with many unfamiliar diagnoses. The critical pathway can serve as an important educational tool, as well as a checklist, so that even patients with less familiar diagnoses do not get substandard care.

In addition, as a greater percentage of ambulatory care is contracted by managed care companies, it is essential that ambulatory care centers are able to cost out their services and document patient outcomes. It is becoming increasingly difficult to contract with managed care companies without first being aware of the costs associated with providing a particular service. As with all insurance companies, managed care companies remain in business by working to keep as much of the patient's premium as possible. They often try to obtain a discounted rate from established fee schedules, negotiate capitation, or reimburse under patient groupings (Daughtery, 1994). It would be fiscally unwise for any ambulatory center to negotiate such contracts, unless there is a high degree of confidence about the actual cost of service. Most managed care companies and many insurance companies are looking to contract with health care providers that have costed out services and consistently use a total quality management approach in monitoring patient outcomes. Therefore, critical pathways not only make good sense in ambulatory settings, but the current climate has made them essential.

DEVELOPING CRITICAL PATHWAYS IN THE AMBULATORY CARE SETTING

The process of developing critical pathways for use in an ambulatory care setting follows the same general principles of development used in other settings. These principal actions are listed in Figure 6.1. For a detailed explanation of the process, see chapter 2.

Keep in mind that pathway development is a continuous process. When used most efficiently, data collected from the critical pathway are continuously fed back into the system to improve patient care, to decrease cost, and to refine the critical pathway. Utilized in this way, the critical pathway is never viewed as finalized, but rather as an evolutionary tool that will become more sophisticated as health care progresses.

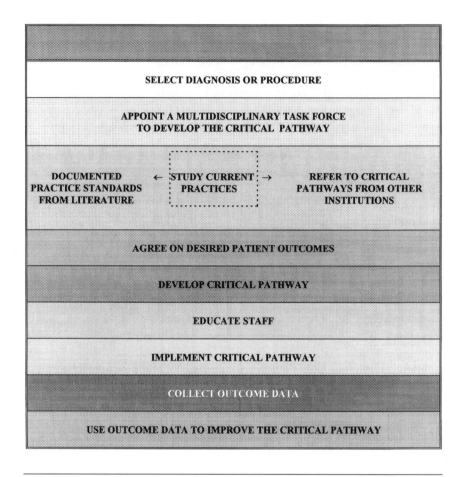

FIGURE 6.1   Steps to critical pathway development and implementation.

Creating an Atmosphere of Collaboration

The key to successful critical pathway implementation in the ambulatory setting is multidisciplinary involvement and commitment. Unless there is a sincere effort to collaborate with all disciplines involved in a particular diagnosis or procedure from the initial stages of planning and writing of the critical pathway, it will be next to impossible to expect collaboration and collegiality during implementation and thereafter. For this reason, the multidisciplinary team should consist of key members of each discipline involved with the diagnosis or procedure under consideration. Multidisciplinary team members should be verbal members of their professional community who will be committed to working out problems voiced within their profession in a constructive, problem-solving manner.

Another key to collaboration is an open, honest discussion of the current health care climate, which has required the emergence of tools such as critical pathways. In the recent past, managed care was relatively uncommon. Today the reality is that without managed care contracts, it would be difficult for many ambulatory care centers to remain solvent. Therefore, ambulatory care centers must document patient outcomes and compete for business by costing out their services. Critical pathways provide a means to track both patient outcomes and cost in an attempt to jockey for a position in the health care marketplace.

There are many software programs now available that allow for breaking down and analyzing procedures or treatments by cost and achievement of patient outcomes. Similar information can also be obtained through the use of a simple spreadsheet and data entry. The data collected from the critical pathway can be analyzed in this manner over time. This data, together with patient outcomes and practitioner cost, can be analyzed and used to motivate practitioners to be both quality and cost conscious. Change is a difficult process for all. By creating an atmosphere of collaboration and by educating the staff on the financial realities of health care, the proper groundwork will be done for successful implementation of critical pathways.

Defining Outcomes

Before actually mapping out interventions on a critical pathway, it is important for the multidisciplinary team to define adequately what patient outcomes are desirable for a particular diagnosis or procedure. By defining patient outcomes and then incorporating them into the critical pathway, interventions can then be mapped out that are specifically designed to facilitate achievement of those outcomes.

In addition to patient outcomes that are diagnosis or procedure specific, patient satisfaction is an increasingly important outcome to track when competing for managed care contracts. Because managed care companies want to keep their enrollment numbers up, evidence of positive patient satisfaction data is frequently required before choosing a preferred provider. It is important that the multidisciplinary team have an ongoing mechanism in place for tracking patient satisfaction, of using that information to create system changes, and to improve patient care where necessary. Including a mechanism that gives the patient feedback on any input gives the patient the message that his or her concerns are taken seriously and may contribute to an overall feeling of patient satisfaction.

Considering Cost Concentration

As mentioned in earlier chapters of this book, there are certain focal points of critical pathway development that are dependent on the site of implementation. Each setting has some general areas that are associated with cost concentration. In ambulatory care these areas include facility costs, the cost of laboratory testing, and the cost of supplies. Often tests are routinely ordered, even if the test result is not linked to patient outcome. Asking the staff to consider the effect of all interventions, including testing on patient outcomes, may have significant influence on what is ultimately put into the critical pathway.

For ambulatory surgery centers, operating room time is an additional area of cost concentration. Collecting data to document the total cost of a procedure, including operating room time, equipment and supplies, and organizing this data by individual staff members or teams, help the staff to see clearly the effect of the care that they deliver on the ultimate cost to the patient and the institution. In addition, many practitioners may not be aware of the differences in cost between various types of equipment or implants used for patients. Educating the staff on these differences is often enough to change behavior and thus lower cost. It is important to make individuals involved in critical pathway development aware of specific areas of cost concentration before beginning to write the critical pathway. This information will provide a reality-based focus when creating a critical pathway that carefully balances quality of care and financial realities.

Writing the Pathway

Once the multidisciplinary team has agreed on desirable patient outcomes associated with the diagnosis or procedure under consideration, they can work back from these outcomes to write the critical pathway. As mentioned in earlier chapters, sample critical pathways can be obtained from a variety of sources, including other ambulatory care centers, hospitals, and health care books and journals. In addition, some pharmaceutical companies have developed critical pathways that they will share through their sales representatives. There are also many consulting firms that have critical pathways for purchase, or that will develop critical pathways to suit individual needs. It is helpful to give the staff a variety of critical path formats to consider so that they may integrate positive elements of many critical pathways into one critical pathway that will best suit their needs.

It is imperative in ambulatory care, as in all other areas, that the critical pathways be developed by the individuals that will work with them. Although there are commonalities among similar types of agencies and organizations, every health care organization has its own personality complete with a specific set of needs. These needs can be met best by the individuals who know the ins and outs of the organization and what makes it work. A set of critical pathways designed to be generic or to meet the needs of one organization may not take into account the distinct needs of other organizations. Sample critical pathways are most useful as comparative tools for the staff to use when developing their own critical pathways. In this manner, the sample critical pathways may be used as a teaching tool to educate the staff about critical pathway format. The sample critical pathway may also serve as a template and a springboard for discussion of staff interventions and patient outcomes when designing critical pathways.

Sample critical pathways or purchased critical pathways may be implemented with the intent of using staff feedback to restructure and reorganize the critical pathway so that is useful to them. It is wise to first present the critical pathway to the staff and give them an opportunity to make initial changes that they believe will make the critical pathway useful in their setting. Content should be carefully considered to be sure that it is not in direct opposition with current practice, or the critical pathway could set the staff and the pathway up for failure. Once implemented, the pathway can be revised on a regular basis, using the data that are collected through documentation, staff feedback, and patient satisfaction surveys. After a period of time, the critical pathway will truly reflect staff input.

SAMPLE CRITICAL PATHWAYS FOR USE IN AMBULATORY CARE

One major distinction of critical pathways written for use in ambulatory care is the time frame. While the time frame for acute care critical pathways is often written in terms of days, critical/emergency care in terms of hours, and home care in terms of visits, the time frame in ambulatory care is often a combination of these, depending on the treatment or procedure under consideration. Some treatments may be completed over a period of days or weeks, while other surgical procedures are completed in a matter of hours. These time frames must be considered when writing the critical pathway.

If all associated assessments, consultations, and pretesting will be done before the day of the procedure, these should still be included on the critical pathway. In addition, any relevant follow-up care should be present. The preadmission education and testing may need to be done in the physician's office. The follow-up care may need to be completed in a clinic, or over the telephone in the patient's home. This must all be considered when designing the critical pathway. For this reason, critical pathways used in ambulatory care often consider a continuum of care, rather than a single episode. Figure 6.2 is a critical pathway created for use with patients receiving outpatient electroconvulsive therapy treatments. Figure 6.3 is a critical pathway created for use in ambulatory surgery. In addition to outlining the care and treatment during the actual procedure, these critical pathways encompass the preadmission work-up and discharge outcomes.

The sample critical pathways in Figures 6.2 and 6.3 are utilized as nursing documentation tools as well as quality improvement monitors. The front sheets of the critical pathways serve as multidisciplinary treatment plans and clearly outline patient problems, treatment goals, nursing diagnoses, and the plan of care. Once the plan of care is reviewed with the patient, the patient signs the treatment plan to document agreement and understanding. Because there are spaces for initiation

Patient Name:_____
ID#:_____

PRE-EXISTING FACTORS:

" age >70	" diabetes mellitus	" hypertension	" psychosis	" homeless

ASSESS/OBSERVE: (Presence requires documentation)	Date Identified/ Initials	COMORBIDITIES/ PATIENT PROBLEMS		Date Resolved/ Initials
"mood disturbance "eating problems "somatic complaints "suicidality "sleep problems "anhedonia "anxiety/agitation "self-mutilation "impaired cognition "thought disorder "hopelessness/helplessness "Other:				

LONG TERM GOALS	RESPONSIBLE DISCIPLINES	TARGET DATE
☐ To improve mood and affect by _____.		
☐ Documented decrease in neurovegetative signs/symptoms of depression by _____.		
☐ Documented decrease in psychotic symptoms by _____.		
☐ Re-establish independence with ADLs, appetite, sleep by _____.		
☐ Other:_____		
☐ Long-Term Personal Goal:_____		

NURSING DIAGNOSES: (Choose appropriate diagnoses)	Resolved/Date Initials
☐ Knowledge deficit regarding ECT and medications administered during procedure.	☐ Yes ☐ No
☐ Knowledge deficit related to rationale for treatments, side effects, and risks associated with ECT.	☐ Yes ☐ No
☐ Anxiety (moderate to severe) related to impending therapy.(Have patient rate anxiety on scale 1-10)	☐ Yes ☐ No _____
☐ Moderate to high risk for injury/aspiration related to ECT and altered level of consciousness immediately following ECT.	☐ Yes ☐ No _____
☐ Altered thought processes related to side effects of temporary memory loss and confusion.	☐ Yes ☐ No
☐ Activity intolerance/self-care deficit related to incapacitation/confusion/memory loss during postictal stage. (circle appropriate choice)	☐ Yes ☐ No _____
☐ Other:_____	☐ Yes ☐ No

PATIENT EDUCATION

TEACHING TOOLS:	PATIENT LEARNING NEEDS:
☐ Patient Critical Pathway	
☐ ECT Video	
☐ ECT Handout	
☐ Other:	

INITIALS/SIGNATURES:

VARIANCE CODES:

V-1 PATIENT RELATED	V-2 CARE PROVIDER RELATED	V-3 SYSTEM RELATED

_____
Signature and Title

X_____
Patient Signature: I have reviewed the above goals
and I am in agreement with the plan of care

FIGURE 6.2a   Critical pathway for outpatient electroconvulsive therapy.

	PRE-ADMISSION	PRE-ECT	ECT	POST-ECT (Until Stable)	WHEN STABLE/ D/C OUTCOMES
CLINICAL ASSESSMENTS	Medicine: □ History & Physical ___ □ Anesthesia Clearance ___ □ Medical Clearance ___ □ Neuro Consult (If organic pathology suspected) ___	Medicine & Nursing: □ Documentation of ECT justification. Nursing: □ Pre-ECT Check List □ Pre-ECT Mental Status ___ (short term memory/confusion) □ Assess Pre-ECT anxiety	Nursing: □ Pulse Oximeter. □ Tolerance of procedure.	□ Vital Signs Q15 minutes. □ Pulse Oximeter until stable.	□ Assess sitting and ambulation . □Demonstrates proper tissue perfusion before/after treatment (absence of cyanosis of severe mental status change)
ACTIVITY/ SAFETY		□ Close Observation		□ Patient to remain on side to maintain airway. □ Allow to sleep/rest.	□ As tolerated. □No aspiration. □Undergoes treatment without sustaining injury. □ Demonstrates ability to ambulate to restroom with minimal assistance.
LABORATORY TESTS/ TREATMENTS	N AB* (Please check) Bloodwork: (within 30 days) □ CBC ___ ___ □ Lytes ___ ___ TESTS: □ EKG ___ ___ □ Urine Analysis ___ ___ □ EEG, CAT Scan ___ ___ (If organic pathology suspected)	□ Written informed consent from patient, family member or responsible party. □ Patient must void.	Document Time of: □ Oxygen. □ Anesthesia. □ Medication. □ ECT. □ EKG leads applied/removed. □ Release from OR and condition . Document : □ IV solution used □ Duration of seizure	□ Document time of return to consciousness. Orient patient to: □ Time □ Place □ Person *Assess for side effects of ECT including: □ Hypotension □ Agitation □ Headache □ Nausea □ Disorientation □ Memory Loss	□Laboratory test results are within normal limits. □Patient will maintain reality orientation following ECT treatment.
MEDICATIONS		□ Obtain orders for required medication .			
DIET		□ NPO .		□ NPO.	□ Fluids . □ Light meal.
DISCHARGE	□ Transportation to /from ECT coordinated				□ Discharge to responsible party.
TEACHING	□ ECT teaching tool to patient. □ ECT Patient Critical Pathway. ___ □ ECT Video. ___	□ Explain purpose of ECT. □ Explain ECT Protocol. □Discuss medications used during ECT and potential SEs.			□ Verbalizes decrease in anxiety level following explanation of procedure and expression of fears.

KEY: *Presence requires documentation; ___ If unmet, record variance code

FIGURE 6.2b Critical pathway for outpatient electroconvulsive therapy.

Patient Name:_____
ID#:_____

PRE-EXISTING FACTORS:

"age>70   "homeless   "hypertension
"diabetes mellitus   "cardiac problem:_____
                              "other:_____

ASSESS/OBSERVE: (Presence requires documentation)	Date Identified/ Initials	COMORBIDITIES/ PATIENT PROBLEMS	Date Resolved/ Initials
" Vital Signs " Pain " Mental Status " Edema " Capillary Refill " Bowel Sounds " Urination " Nausea/Vomiting " Bleeding "changes in skin color (pallor, redness, cyanosis) "Other:			

GOALS	RESPONSIBLE DISCIPLINES	TARGET DATE
☐ Adequate knowledge of follow-up care, as evidenced by client's ability to describe the condition, risk factors, reportable signs and symptoms, required life-style adaptation and follow-up care.		
☐ Absence of signs and symptoms of complications related to procedure.		
☐ Absence of pain or level of pain reported as tolerable by patient.		
☐ Other:_____		
☐ Long-Term Personal Goal:_____		

NURSING DIAGNOSES: (Choose appropriate diagnoses)	Resolved/Date Initials
☐ Knowledge deficit regarding procedure and associated mediations.	☐ Yes ☐ No
☐ Knowledge deficit related to rationale for treatments,side effects, and risks associated with procedure.	☐ Yes ☐ No _____
☐ Anxiety (moderate to severe) related to impending surgery.(Have patient rate anxiety on scale 1-10)	☐ Yes ☐ No _____
☐ Moderate to high risk for injury/aspiration related to procedure.	☐ Yes ☐ No
☐ Other:_____	☐ Yes ☐ No

PATIENT EDUCATION	
TEACHING TOOLS:	PATIENT LEARNING NEEDS:
☐ Patient Critical Pathway	
☐ Preoperative Video (Title:_____)	
☐ Patient Handout (Title: _____)	
☐ Other:	

INITIALS/SIGNATURES:

VARIANCE CODES:		
V-1 PATIENT RELATED	V-2 CARE PROVIDER RELATED	V-3 SYSTEM RELATED

_____   X _____
Signature and Title   Patient Signature: I have reviewed the above goals
                                       and I am in agreement with the plan of care

FIGURE 6.3a   Critical pathway for outpatient surgery.

	PRE-ADMISSION	PRE-OP	OPERATIVE PHASE	POST-OP (PACU Until Stable)	WHEN STABLE/ D/C OUTCOMES
CLINICAL ASSESSMENTS	**Medicine:** ☐ History & Physical ___ ☐ Anesthesia Clearance ___ ☐ Medical Clearance ___ ☐ All appropriate consults obtained (list) ___ ☐ Other ___	**Medicine & Nursing:** ☐ Documentation of surgical justification. **Nursing:** ☐ Pre-op Check List ☐ Assess Pre-op anxiety	**Nursing:** ☐ Support & maintain body alignment. ☐ Maintain skin integrity. ☐ Tolerance of procedure ☐ Other: ___	*Assess for side effects of anesthesia including: ☐ Hypotension ☐ Agitation ☐ Headache ☐ Nausea ☐ Disorientation ☐ Other: ___ ☐ Document time of return to consciousness.	☐Nursing: ☐Demonstrates proper tissue perfusion before/after surgery (absence of cyanosis of severe mental status change). ___ ☐Demonstrates normal breathing pattern & airway clear as evidenced by normal respiratory pattern, patent airway, and decreased or absent adventitious breath sounds. ___
ACTIVITY/ SAFETY	☐ Ad lib ☐ Other:	☐ Ad lib	☐ Bedrest	☐ Bedrest	☐ Assess sitting and ambulation ☐ Demonstrates ability to ambulate to restroom with minimal assistance
LABORATORY TESTS/ TREATMENTS	N AB* **Bloodwork:** (Please check) (within 30 days) ☐ CBC ___ ___ ☐ Lytes ___ ___ ☐ Other: ___ **TESTS:** ☐ EKG ___ ___ ☐ Urine Analysis ___ ___ ☐ Chest X-Ray ___ ___ ☐ Other: ___	☐ Repeat abnormal lab results ___ ☐ Written informed consent from patient, family member or responsible party. ☐ Shower ☐ Empty bladder ☐ Start IV	**Tests:** ☐ C&S ☐ Other: ___ **Document Time of:** ☐ Oxygen. ☐ Anesthesia. ☐ Medication. ☐ Release from OR and condition.	☐ Vital Signs Q15 minutes. ☐ Pulse Oximeter until stable. ☐ Cardiac monitoring ☐ O2 ☐ IV: ___ ☐ Other: ___ **Orient patient to:** ☐ Time ☐ Place ☐ Person	☐ Laboratory test results are within normal limits. ☐ Exhibits stable vital signs for at least 30 minutes prior to discharge. ___ ☐ Is free of complications related to surgical procedure; excessive bleeding, neurovascular impairment, edema, urinary retention. ___
MEDICATIONS		☐ Obtain orders for required medication ☐ Pre-op meds ☐ Prophylactic IV antibiotics	☐ Per anesthesia	☐ Medicate for pain: ___ ☐ Medicate for nausea: ___	☐ Verbalizes acceptable pain control. ___
DIET	☐ Regular ☐ Other:	☐ NPO.	☐ NPO.	☐ NPO.	☐ Tolerates fluids. ___ ☐ Tolerates light meal.
DISCHARGE	☐ Transportation to/from procedure coordinated ___	☐ Address any D/C needs ☐ Verify D/C transportation			☐ Discharge to responsible party. ___ ☐ Patient/SO verbalize understanding of d/c instructions including self-care activities, medications and treatments and reportable S&S.
TEACHING	☐ Teaching tool to patient. ☐ Patient Critical Pathway ___ ☐ Pre-op Video ___	☐ Pre-op instructions ☐ Explain surgical protocol. ☐ Discuss medications used during/immediately after procedure and potential SEs		☐ Reinforce coughing/deep breathing exercises. ___	☐ Demonstrates effective coughing/deep breathing. ___ ☐ Other: ___

KEY: *Presence requires documentation; ___ If unmet, record variance code

FIGURE 6.3b Critical pathway for outpatient surgery.

87

and resolution of all goals, problems, and nursing diagnoses, only those unresolved at the time of discharge need to be addressed in a narrative note. This format allows for easy reference prior to future treatments and concise collection of variance data, without requiring an additional documentation tool. The ambulatory surgery critical pathway can be used as a template and be made diagnosis specific for each surgical procedure, or it can be left general to cover all surgical procedures done in an ambulatory setting.

AMBULATORY CARE AND THE FUTURE

As the health care delivery system in the United States continues to evolve, experts agree that more care will be delivered in outpatient settings in the future. It is imperative that ambulatory care centers prepare for this change by carefully considering the way they deliver care and the way they do business. Critical pathways provide a means of considering both clinical and financial aspects of care in an objective, patient-focused manner.

Critical pathways used in the ambulatory setting tend to be more progressive than the critical pathways found in other areas, in that they often address a continuum of care, rather than a single admission or episode of care. As health care continues to extend to the outpatient setting, critical pathways of ambulatory care will need to address other stops along the continuum. This enables the provider to keep less frequented settings in focus and to utilize all resources of the health care system to their potential.

REFERENCES

Daughtery, J. (Ed.). (1994). Managed care companies are at your door-Do you know what your costs are? *Same-Day Surgery, 18*(9), 113–116.

Howe, R. (1996). *Clinical Pathways for Ambulatory Care Case Management.* Gaithersburg, MD: Aspen.

National Center for Health Statistics (1992). *Health, United States* (p. 192). Hyattsville, MD: Public Health Service.

BIBLIOGRAPHY

Daughtery, J. (Ed.). (1992). LOS reduced one hour with critical pathways. *Same-Day Surgery, 16*(10), 150–153.

Daughtery, J. (Ed.). (1992). Warning: APGs may lead to managed care. *Same-Day Surgery, 16*(5), 76–79.

Daughtery, J. (Ed.). (1994). The latest on critical pathways: Intra-op, post-op phases added. *Same-Day Surgery, 18*(11), 145–147.

Daughtery, J. (Ed.). (1994). Same-day surgery managers prepare for onslaught of managed care, explosion in outpatient surgery. *Same-Day Surgery, 18*(1), 1–3.

Daughtery, J. (Ed.). (1995). ASCs seek to win managed care contracts. *Same-Day Surgery, 19*(2), 22–23.

Norman, L. (1995). Computer-assisted quality improvement in an ambulatory care setting: A follow-up report. *Joint Commission Journal on Quality Improvement, 21*(3) 116–131.

Simkin, B. (1995). Transitional Pathway encompasses outpatient settings. *Critical Path Network, (1)*, 9–17.

7

Critical Pathways in Home Care: Developing a Psychiatric Home Care Program

Kathleen Wheeler

HOME CARE ECONOMICS

The health care system has changed dramatically over the past five years. As a result of the growth of managed care, rising costs, and public interest in health care, a new age of accountability and cost-effectiveness has emerged. The current health care climate mandates that providers deliver the best treatment with the maximum resources in the shortest amount of time. Thus, outcome driven care has become a priority for all providers.

The need to curb costs has contributed to the dramatic growth of home care nursing as hospitals downsize and more patients are discharged earlier and sicker from inpatient settings. Home care is the most rapidly growing sector of the health care system. With the continued growth of the elderly population, more home care and community-based services will be needed in the future (Dee-Kelly, Heller, & Sibley, 1994).

The reimbursement system for home care is late to change from retrospective reimbursement to prospective payment, and this has added to the impetus to curb costs while maintaining quality care. Although some services continue to be paid on a fee-for-service basis, other models are being developed with a greater emphasis on cost containment. The traditional "fee-for-service" model for home care services provided more services to achieve clinical outcomes and thus made more dollars for the agencies. As home care progresses toward prospective payment, nurses are asked to accomplish more with less.

Managed care companies are now restricting the number of nurse visits and requiring precertification for the nurse to provide service for a specified time period. A predetermined number of visits is authorized per episode of illness. The nurse is frequently justifying care, and spending time and energy convincing the case managers of managed care companies of the need for services. Besides the increase in the number of people who belong to managed care companies, some state governments are asking managed care companies to provide care for both Medicare and Medicaid recipients.

GETTING STARTED

Case Management and Critical Pathways

The need for a psychiatric program was identified by a large home care agency in southern Connecticut. The agency's staff nurses noted that many patients in the community with medical problems were not being discharged by the home care agency because of emotional problems which were interfering with their ability to learn about or to manage their illnesses. These included memory problems, delusional thinking, family relationship problems, eating problems, sleeping problems, anxiety, somatic complaints with no etiology, obsessive thinking or worrying, feelings of guilt, worthlessness, hopelessness, indecisiveness, and tearfulness, noncompliance with medications and/or treatment, alcohol or substance abuse, suicidal thoughts, demanding and angry behaviors, and pain management problems. Elderly patients, who are the largest population of home care patients, were identified as being at especially high risk. Emotional and cognitive changes which are thought to be a normal part of aging may actually indicate depression, which can be successfully treated 90% of the time if accurately diagnosed.

The care delivery system at the agency is a case management system with individualized care plans for each client. Maintaining the patient in the community using a case management model reduces the need for hospitalization and saves money. Case management is a logical extension of the nursing process with intake and assessment, development of a plan of care, implementation of the plan, documentation and supervision of ancillary health care providers such as home health aides, coordination of care and discharge planning, and referral if necessary. Each patient admitted to the agency is assigned a Primary Nurse Case Manager who is a registered nurse responsible for case management throughout the patient's course of care. The registered nurses are usually assigned to a specific geographic area and are supervised by a Clinical Nurse Manager.

However, psychiatric patients frequently present complex case management challenges. Often patients have multiple stressors, inadequate social support, are isolated, and, because of transportation or physical problems, are either unable or unwilling to access community resources for psychotherapy.

The primary goal for the home care nurse is to assist the individual toward maintaining independence in the community. This is accomplished through linking patients and caregivers to outpatient community resources and coordinating services such as pharmacy, social service, and home health aide services as well as assessment and treatment of mental health problems in order to restore or maintain functioning.

Critical pathways are an excellent tool to implement a case management system of care. Although originally developed for inpatient acute care settings, critical pathways have expanded to other health care settings such as ambulatory care settings, rehabilitation, long-term care settings, and home care settings. Home care has been late to develop critical pathways because of the slower impact of managed care on reimbursement issues for home care. One recent publication did address critical pathways for home care clients with congestive heart failure (Cooke & Brodrick, 1994).

Critical paths are now being developed for mental health problems and for mental health settings. Although efforts to quantify behavioral or mental-health outcomes through research are only just beginning, a preliminary study reported that after initiation of critical paths, the average length of stay decreased 8 days over a 12-month period in an inpatient psychiatric hospital (Dunn, Rodriguez, & Novak, 1994). Another report cited a decreased length of stay by nearly 4 days for 5 hospitals using critical paths for psychiatric inpatients (Homan, 1994). There are no reported studies of outcomes for psychiatric home care utilizing critical paths.

Policy Development

Starting a psychiatric home care program in an already accredited home care agency is easier than starting home care services as part of a hospital or other mental health agency because of the many regulatory and accreditation bodies home care must answer to (JCAHO, CHAP, HCFA, NAHC, CON, DHHS). Since the psychiatric home care program was to be an addition to an already well-established home health agency, the Policy and Procedure Manual for the agency became a starting point for developing the program in order to establish regulatory compliance and also to ensure consistency in philosophy, policy development, and documentation. Thus the agency's policies were adapted for the psychiatric program with an effort to utilize admission, discharge, and documentation procedures already in place. Policy development became an ongoing process as the community's and agency's needs were identified.

A nursing care plan model of care was used at the agency with each primary nurse developing the care plan for her patient. Besides being time-consuming to write, many interventions fell through the cracks. If the primary nurse was absent and another nurse visited the patient, it was difficult to tell exactly where the patient was in terms of their care. Without specific protocols for each visit, care often becomes fragmented, and it is difficult for new or part-time staff to give care that is consistent and comprehensive. For psychiatric patients, communication, consistency and continuity of care are crucial for positive outcomes. Critical pathways address the need for an efficient, comprehensive plan of care and become the vehicle for the implementation of the psychiatric program and a tool to guide psychiatric home care practice.

DEVELOPING THE PATHWAYS

Steps in developing critical pathways include:

- agency administration commitment to the process
- identification of high-volume mental-health problems
- library research of nursing and psychiatric literature for specific problems/diagnoses
- delineation of outcomes and interventions
- determination of time frames and visit frequency
- evaluation, development and/or procurement of assessment, teaching, and outcome tools
- orientation of nurses to the use of critical pathways
- feedback from nurses and kinks ironed out

Since the program was new to the agency, it was not possible to review past agency records in order to determine practice patterns for past psychiatric problems. Client mental-health problems usually were undiagnosed or untreated. A mental-health consultation system was established within the agency for the nurses, and team meetings were held to introduce the program and to learn from the nurses what kinds of patient problems the psychiatric nurses might be able to help with. As mental-health problems were identified, consultations increased. Inservices were held with the agency nurses explaining the psych program, and case presentations were given to illustrate successful interventions and collaboration. A memo was distributed to all staff nurses explaining our service and listing patient problems the psychiatric program could help with.

The Psychiatric Home Care Program that evolved includes these services: geriatric mental-health assessment and treatment, mental-health consultation to meet the needs of the homebound medically ill patient, psychiatric case management for patients with serious and persistent mental illness, anxiety and phobia treatment, and home detoxification. Those eligible for these services include the homebound who have a primary medical diagnosis with psychiatric needs or who are diagnosed with a psychiatric disorder, are under the care of a physician, and require the skilled services of a psychiatric nurse.

It became apparent early in program development that many of the problems patients faced revolved around issues of depression or anxiety. These were the first two critical pathways to be developed. As the program expanded to include persons with a primary psychiatric disorder, schizophrenia and home detoxification critical pathways were also developed.

The first step in developing critical pathways was to identify desirable outcomes with specific time frames and then to map out key interventions for each visit to achieve those outcomes. This was accomplished through a library literature review of the research for the problem/diagnosis identified, the A.N.A. Standards of Care, agency policies, and consultation with Clinical Specialists in Psychiatric Nursing. Several templates of critical pathways which were already developed were obtained and used as models (VNA First, 1994; Gingerich & Ondeck, 1996).

Most of the critical pathways are structured over a 14 visit trajectory with a discharge visit. Visits are usually 3 times a week for the initial assessment period and then twice a week depending on the severity of the patient's problem. Most of the initial pathways were piloted for a 2-month period, the certfication period for Medicare reimbursement.

Often the psychiatric nurses "share" patients with the medical nurses in order to assist the medical nurse in ongoing psychiatric assessment and treatment when necessary. Visit frequency and time of service for each episode of illness are individualized and discussed in clinical conferences. Sometimes only a comprehensive mental-health assessment is indicated in order to determine placement issues or to help the medical nurse plan care for her client.

In determining the frequency of visits, consideration was given to the problem of negative reinforcement for positive gains due to withdrawal of services when the client begins to make progress. Clients often develop dependent relationships with their nurses and sometimes the nurse is the only person the client sees that week. If visits are decreased as clients improve, frequently symptoms increase. For this reason attempts are made to keep visit frequency consistent after the initial assessment period and throughout noncrisis times. Linking clients to community services is initiated early on and helps to assuage dependency needs so clients do not feel abandoned as symptoms ameliorate. Nurses also initiate telephone calls to clients during the week on days when visits are not scheduled in order to stabilize clients who are identified as needing this contact.

Each visit on the critical pathway includes care elements which are general categories of care such as medication, nutrition/hydration, tests, activity, safety, treatments, and interteam services/community referrals. Within each care element category, interventions are delineated to accomplish the identified patient outcomes. See Figure 7.1 for an example of the Depression Critical Pathway. The critical pathways are written in phases which allows the nurse to evaluate the patient's progress based on predetermined patient outcomes, rather than on an exact number of visits. The initial phase is the assessment and planning phase. Generally the designated patient outcomes for this phase are completed in one to three visits.

Patient Name:_____

Medical Record #:_____Date:_____

Patient/family member primary language:_____

Assessment Completed	Score (Where Indicated)	Date Initiated	Comorbidities, Problems or Other Copathways	Date Resolved
☐ Psychiatric Assessment Intake:				
☐ Beck Depression Inventory:				
☐ GAF:	Admission:			
	Highest in past year (If Known):			
☐ Geriatric Depression Scale:				
☐ Patient/Family Status:	Knowledge:			
	Behavior/Skill:			
	Psychosocial:			
	Physical:			

NURSING DIAGNOSES: (Choose appropriate diagnoses)

☐ 1. Knowledge deficit regarding disease process and home care management, and medication.

☐ 2. Depression related to situational/maturational crisis.

☐ 3. Self-esteem disturbance related to feelings of abandonment, numerous failures, punitive superego, impaired cognition.

☐ 4. Social isolation related to fear of rejection, unresolved grief, absence of available significant other/ peers (circle).

☐ 5. Ineffective individual/family coping related to diagnosis (circle appropriate problem).

☐ 6. Altered nutrition related to anorexia.

☐ 7. Other:

LONG TERM GOALS	RESPONSIBLE DISCIPLINES	TARGET DATE	REVIEW DATE
☐ To identify and mourn losses.			
☐ To develop coping skills.			
☐ To identify and utilize support available in community or from significant other			
☐ To exhibit increased feelings of self-worth as evidenced by verbal expression of positive aspects about self, past accomplishments, and future prospects.			
☐ Patient/family member long-term goal:_____			
☐ Patient/family member primary concern:_____			
☐ Other:			

TEACHING TOOLS:
☐Weekly plan for recovery of depression
☐Beck Inventory
☐Geriatric Depression Scale
☐ Other (please list):_____

OTHER DISCIPLINES/SIGNATURES
_____
_____
_____
_____

SN VISIT FREQUENCY:
Recommended: 3 wk x 1, 2 wk x 5 (6-13 visits total)
    Assessment /Planning Phase: 1-3 Visits
    Treatment Phase: 4-7 Visits
    Discharge Phase: 1-3 Visits

Ordered:_____

_____   _____
Signature and Title   Date

X_____
Patient Signature/Date: My nurse has reviewed the plan of care with me and I am in agreement with the stated goals.

FIGURE 7.1a   Critical pathway for depression: patient treatment plan.

Patient Name: _____

Assessment/Planning Phase: Visit_____   **Medical Record #:**_____   **Date:**_____
Vital Signs: BP_____   P: R/A   ( )Reg ( )Irreg   R_____   T_____

Care Elements	Interventions: Use "√" for complete; "ND" for not done; "0" for not present.	Comments
DISEASE PROCESS	Complete SNV Psychiatric Assessment Report. __ Assess knowledge of disease process, signs and symptoms of complications, and rehabilitation process. __ Introduce disease process including: Definition of depression. __ * Causes of depression.___ * Reportable Signs and Symptoms.__ Obtain recall on instruction completed during previous visits. __ * *Individualize for patient.*	☐ New Problem / ☐ Change in Status
MEDICATION	Review medication schedule. __Instruct purpose, action, side effects of following medication(s): _____ Set up medi-planner if appropriate. _____	
NUTRITION/ HYDRATION	Assess diet history for adequate nutrition. ___ Assess for and treat symptoms of dehydration. ___	
ACTIVITY	Home bound status: cane, __ walker, __ 0_2, __ Other:_____ Assess functional status. ___ Discuss benefits of regular exercise program. ___	
SAFETY	Assess environmental safety factors and remove barriers as appropriate. ___Assess understanding of how and when to call for help. ___ Assess for suicidality. ___	
TREATMENT	Assist in identifying feelings. ___ Encourage verbalization of feelings. ___ Develop weekly plan for recovery from depression. ___ Instruct regarding: management of depression. ___ Assist patient in identifying losses. ___ Set goals for upcoming week. ___	List goal # from weekly plan: 1.) _____ 2.) _____ 3.) _____
TESTS	Laboratory tests as ordered. ___ If patient on Lithium, establish protocol for Lithium levels. ___	
PSYCHO/ SOCIAL	Assess family/social supports. ___ Begin development of therapeutic relationship. ___ Assess pt./family perception of primary problems.__Assess pt./significant other coping skills. ___ Provide psychotherapeutic counseling. ___ Offer supportive services as appropriate. __ Define expectations of home care including: use of clinical pathway ___, role of the patient ___, role of significant other ___, role of home care staff.___ Complete MSQ. __Assess using either Beck Depression Inventory ___ or Geriatric Depression Scale. ___	
INTERTEAM SERVICES/ COMMUNITY REFERRALS	Assess need for community resources__ written visit schedule.__ Next MD appointment(date). _____ **Supervision of:** HHA ( ) HM ( ) Other ( ) Name/Title:_____New Orient ( ) HHA/ HM following plan of care ( ) Goals being met ( ) **Conference with:**PT ( ) OT ( ) ST ( ) HHA ( ) MSW ( ) MD ( ) Supervisor ( ) Other ( ): _____ Review plan of care for home care services.__	

MD Contact/Outcome:_____

Patient Goal/Plan For the Next Visit:_____**Date of next visit:**_____

_____   ☐ **Plan of care reviewed with patient and patient verbalized**
Signature and Title   **agreement with plan and stated goals.**

FIGURE 7.1b   Critical pathway for depression.

Patient Name:

Treatment Phase: Visit_____ **Medical Record #:_____** **Date:_____**

Vital Signs: BP_____ P: R/A_____ ( )Reg ( )Irreg R_____ T_____

Care Elements	Interventions: Use "√" for complete; "ND" for not done; "0" for not present.	Comments
DISEASE PROCESS	Assess general condition via interview/SNV Psychiatric Assessment Report. ___ Obtain recall on instruction completed during previous visit. ___ Weight _____Stable( )Gain ( ) _____ Loss( ) N/A ( ) Pain:_____ N/A ( )	☐ New Problem / ☐ Change in Status
MEDICATION	Review medication schedule. ___Instruct purpose, action, side effects of following medication(s): _____	
NUTRITION/ HYDRATION	Assess nutritional and hydrational status. ___Review diet restrictions. ___Ask for 24 hour diet recall to assess compliance and understanding. ___Review 24 hour diet recall. ___	
ACTIVITY	Home bound status: cane, __ walker, __ 0$_2$, __ Other:_____ Review role of exercise in symptom management. ___ Develop regular exercise program. ___ Instruct patient to record exercise on weekly plan. ___	
SAFETY	Assess for suicidality. ___Continue to assess environment for risk factors. __ Review emergency procedures and phone #s. ___	
TREATMENT	Assess progress with weekly plan. ___ Review plan/goals from previous visit. ___ Assist in identifying___ and mourning past losses. ___ Begin thought diary.__Instruct regarding suppressed feelings. __Assist with identifying symptoms of suppressed feelings___ Encourage verbalization of feelings. ___ Assist patient with identifying areas of strength/self-worth. ___ Instruct regarding cognitive strategies. ___ (See weekly plan) Instruct regarding strategies for management of depression. ___ Instruct regarding strategies/long-term plan for recovery. ___ Assist in identifying negative or irrational thoughts. ___ Begin thought stopping and substitution. ___ Assist patient in identifying strengths and assets. ___ Develop a plan to express emotions as they arise. ___ Set goals for upcoming week. ___Assess with depression scale (Visit 8 or half-way through treatment phase)___	List goal # from weekly plan: 1.) _____ 2.) _____ 3.) _____
TESTS	Laboratory tests as ordered. ___	
PSYCHO/ SOCIAL	Continue to assess family/social supports___ Continue to assess pt./significant other coping skills. __ Provide psychotherapeutic counseling. __ Assist in identifying significant others for support. ___	
INTERTEAM SERVICES/ COMMUNITY REFERRALS	Identify support groups in community. __ Next MD appointment(date). _____ **Supervision of:** HHA ( ) HM ( ) Other ( ) _____ New Orient ( ) HHA/ HM following plan of care ( ) Goals being met ( ) **Conference with:**PT ( ) OT ( ) ST ( )HHA( ) MSW ( ) MD ( ) Supervisor ( ) Other ( ): _____	

MD Contact/Outcome:_____

Patient Goal/Plan For the Next Visit:_____**Date of next visit:_____**

_____   ☐ Plan of care reviewed with patient and patient verbalized
Signature and Title   agreement with plan and stated goals.

FIGURE 7.1c Critical pathway for depression.

		Patient Name:_____

Discharge Phase: Visit_____   **Medical Record #:**_____**Date:**_____
Vital Signs: BP   P: R/A   ( )Reg ( )Irreg   R   T

Care Elements	Interventions: Use "√" for complete; "ND" for not done; "0" for not present.	Comments
DISEASE PROCESS	Assess general condition via interview. ___ Obtain recall on instruction completed during previous visit. ___ Complete SNV Psychiatric Assessment Report. ___ Weight _____Stable( )Gain ( ) _____ Loss( ) N/A ( ) Pain: N/A ( )	☐ New Problem / ☐ Change in Status
MEDICATION	Review medication schedule. ___Instruct purpose, action, side effects of following medication(s): _____	
NUTRITION/ HYDRATION	Assess compliance with dietary restrictions. ___	
ACTIVITY	Home bound status: cane, ___ walker, ___ 0₂, ___ Other:_____ Assess functional status. ___ Discuss benefits of regular exercise program. ___	
SAFETY	Assess environmental safety factors and remove barriers as appropriate. ___ Assess knowledge of emergency measures/phone #s. ___	
TREATMENT	Assess progress with weekly plan. ___ Review plan/goals from previous visits. ___ Continue to reinforce strategies for long-term recovery. ___ Assist in mourning past losses. ___ Discuss termination in 1 - 2 visits. ___	List goal # from weekly plan: 1.) _____ 2.) _____ 3.) _____
TESTS	Laboratory tests as ordered. ___ Establish protocol for follow-up lab tests. ___	
PSYCHO/ SOCIAL	Provide psychotherapeutic counseling. ___ Encourage verbalization of feelings. ___ Assess coping skills. ___ Complete GAF. ___ Assess with MSQ. ___ Assess with depression scale. ___ Encourage to verbalize feelings. ___	
INTERTEAM SERVICES/ COMMUNITY REFERRALS	Consult with other team members about possible d/c in 1 - 2 visits. ___Instruct in availability of community resources. ___ Review plans to discharge from home care services. ___ Refer to community support group of interest. ___ Encourage patient to attend one support group before d/c. ___ Review importance of follow-up visits with physician or therapist. ___Reinforce compliance with treatment after d/c. ___ Next MD appointment(date). _____ **Supervision of:** HHA ( ) HM ( ) Other ( )_____ New Orient ( ) HHA/ HM following plan of care ( ) Goals being met ( ) **Conference with:**PT ( ) OT ( ) ST ( ) HHA ( ) MSW ( ) MD ( ) Supervisor ( ) Other ( ): _____	

MD Contact/Outcome:_____

Patient Goal/Plan For the Next Visit:_____**Date of next visit:**_____

_____   ☐ **Plan of care reviewed with patient and patient verbalized**
Signature and Title   **agreement with plan and stated goals.**

FIGURE 7.1d   Critical pathway for depression.

Note: Used with permission of United Home Care, Inc., Fairfield, CT.

Next is the treatment phase which generally requires four to seven visits to complete the outcomes. The final phase is the discharge phase and outcomes defined in this phase generally require an additional one to three visits. The final visit is a discharge visit, which covers any generic teaching required of patients prior to discharge as well as content specific to the patient's diagnosis. The treatment plan is outlined on the first page of the critical pathway.

A charting-by-exception format is used where the nurse initials the completed interventions right on the critical pathway. Space for comments allows for individualization of care so that additional nursing interventions and other findings may be written directly on the form. Variances occur when the care is unable to be given as stated or patients do not meet predetermined outcomes and as such are noted on the critical pathway. Variances are either due to the patient, the caregiver, or the system and noted as such. A separate outcome monitor sheet keeps track of all patient outcomes and variances allowing the nurse doing the visit to note variances as they occur and to take corrective action when necessary. This format also allows the nurse making the subsequent visit to see at a glance what outcomes need to be given top priority. Unless a variance or problem occurs during a visit, a narrative is not needed, thus saving documentation and paperwork.

Comprehensive Assessment

The psychiatric nurse always completes a comprehensive psychiatric assessment on the initial visit to the patient. The American Nurses Association standards of care for psychiatric nursing practice emphasize that the assessment interview be systematic, comprehensive, and continuous. "Pertinent data are collected from multiple sources using various assessment techniques and standardized instruments as appropriate.The database is synthesized, prioritized and documented in a retrievable form" (ANA, 1994, p.10). Because the psychiatric home care nurse functions autonomously and relies on assessment data to determine care, completeness and accuracy are essential.

Depending on the patient's level of functioning and cooperation, the intake assessment may take two visits. The assessment begins with asking the patient what he perceives his problem to be and how he feels he can be helped. Also included are general information questions regarding the patient's areas of strength, health history, habits, recent changes and support system, vital signs, current medications and client's understanding and compliance with medications, nutritional and elimination assessment, ADL assessment, substance abuse assessment, neurological assessment, and mental status examination including suicidality and /or homicidality. The Global Assessment of Functioning Scale (GAF) rating is completed at this time and again at discharge as well as a patient/family status rating, which is a state-wide rating scale used for all home care patients in Connecticut.

From this data base, the psychiatric nurse determines which critical path would be most appropriate for the patient. Sometimes there is not a clear presenting symptom but mixed features such as a depression with anxiety or an anxiety disorder with depressive features. The most salient symptom determines which pathway is chosen.

Assessment is ongoing throughout the time of service. For each visit, a one page Psychiatric Assessment Report rates the mental and functional status of the patient on a checklist. Further assessment tools appropriate for the symptom or diagnosis are included in the first visit, midway in the visit trajectory, and again at discharge. If warranted, as with the confused or disoriented patient, mental status assessment using the Folstein or Mental Status Questionnaire (MSQ) is administered at every visit. Each critical pathway lists on the first page assessment and teaching tools needed for that problem, and time frames for administering these are written on the appropriate visit sheet. In this way accurate evaluation of cognition or lack of progress with the treatment plan is tracked and documented on the critical path to provide immediate feedback on the patient's status and whether different approaches or interventions are necessary.

In addition to teaching handouts and assessment tools for each diagnosis, each critical pathway is individualized by use of a weekly plan where specific goals are further identified from a menu of interventions for that particular diagnosis or problem. Space is also provided for other individual goals which may not be listed on the weekly plan. See Figure 7.2 for Weekly Plan for Depression.

Specific interventions for each problem are stated in every visit. These may include patient-outcome oriented psychotherapeutic interventions such as to assist the patient in identifying losses, encourage verbalization of feelings, begin a thought diary, assist in identifying negative or irrational thoughts, assist patient in identifying strengths, keep a food diary, as well as many other strategies to promote mental health. Other important interventions include education about medication, modifying medication regimen if necessary and monitoring side effects, therapeutic effectiveness and medication interactions, teaching about diet, exercise, mental illness and mental health, community resources, relaxation and imagery, and so on. These interventions operationalize techniques used in patient education, bereavement counseling, reminiscence therapy, cognitive and behavioral therapy, stress management, crisis intervention as well as other short-term therapies.

Since many home care patients have complex problems with many comorbidities, the need for an auxiliary pathway to address secondary problems was apparent. The pathways developed to address this need are called copaths and form an abbreviated critical pathway. Copaths are one page with a goal and a list of patient-oriented outcomes to be achieved. Per visit interventions are not delineated as they are for critical pathways but weekly plans with individualized goals, assessment, interventions, and teaching tools are utilized and can be implemented along with the client's primary critical pathway. See Figure 7.3 for a list of copaths which have been developed.

The primary admitting diagnosis or the most salient symptom determines the critical path used, and the copath represents the comorbidity of the patient For example, the patient may be admitted to home care for a primary diagnosis of COPD but have a great deal of anxiety which significantly impacts on the patient's health status. This patient is placed on the COPD Critical Pathway and the Anxiety Copath. Conversely, the patient may have a primary psych diagnosis and a secondary medical problem. For this patient, the copath would be medical and the critical path would be psychiatric.

A psychiatric critical pathway may also be combined with a psychiatric copath. For example, a bipolar patient who has a manic phase may be placed on the Depression Critical Pathway with the Manic Copath. Another example of combining a psychiatric critical pathway with a psychiatric copath is a patient with an anxiety disorder who is agoraphobic. For this client, the Anxiety Critical Path is used with the Phobia Copath. The goal for the Phobia Copath is to assist the client in managing anxiety in order to decrease phobic behavior. The interventions for this copath are to be implemented in order because they are sequential steps for in vivo desensitization.

Another copath with sequential steps is the Noncompliance Copath which lists steps to set up a contingency contract. At the top of the copaths instructions which have sequential steps are given to the nurse. Not all copaths require this step-by-step process in giving care. See chapter 3 for examples and more information about copathways.

A further example of using a psychiatric critical pathway with a psychiatric copath is the dual diagnosis patient. For example, if the patient is actively using substances, the Detox Critical Pathway has priority. Or if the dual diagnosis patient is admitted because of a recent discharge from the hospital for a psychotic episode for Schizophrenia, the patient is placed on the Schizophrenic Critical Pathway and the Relapse Prevention Copath. Again, the most salient problem determines which critical path and copath the patient is placed on.

As the need arises, copaths are developed so the program is tailor made for the patient's needs. The psychiatric nurses who care for the patients develop copaths and share them with their colleagues in weekly case conferences. For example, a client of one of the nurses had been placed on the Depression Critical Path and had a crisis which created variances. Subsequently the nurse made a copath for crisis intervention.

The client becomes a partner in care through use of critical pathways by outlining the care that will be delivered and his or her expectations as a partner in that care. Identification of specific outcomes the patient should achieve within a set number of visits assists in coordination and monitoring of changes in care which may be necessary. This model of shared responsibility is particularly important in the home care setting. Since nurses are guests in the patient's home, self responsibility and involvement are imperative for the patient. It is in the home that clients are most receptive, and obstacles to learning and compliance are witnessed firsthand. This gives an immediacy to care that is not otherwise possible. The critical pathway empowers the client and promotes independence.

Patient's Name: _____

Goals for week:
1. _____
2. _____
3. _____

Week of: _____

	MON	TUES	WED	THUR	FRI	SAT	SUN
COGNITIVE INTERVENTIONS							
1. Keeping thought diary.							
2. Made list of positive accomplishments/attributes.							
3. Underlined negative thoughts and said out loud.							
4. Stopped thoughts and substituted positives.							
5. Identified areas of life that are not within ability to control and areas that are within ability to control.							
6. Set realistic goals and prioritized.							
7. Took steps to achieve goals.							
BEHAVIORAL INTERVENTIONS							
8. Got dressed and combed hair.							
9. Stayed out of bed all day.							
10. Made a list of activities enjoyed.							
11. Did pleasurable activity.							
12. Participated in social event.							
13. Talked to a friend.							

FIGURE 7.2   Weekly plan for recovery from depression.

14. Keeping food diary.							
15. Improved nutrition.							
16. Did one-half hour exercise.							
17. Used deep breathing and imagery.							
18. Imagery and visualization of success/positive outcome.							
19. Keeping feeling journal.							
20. Identified/expressed feelings.							
21. Mourned past losses.							
22. Practiced assertive communication.							
23. Nurtured self.							
24. Used medication appropriately.							
25. Read about depression							
26. Utilized spiritual beliefs and practices.							
27. Reached out to others in community.							

FIGURE 7.2 *(continued)*.

Note: Used with permission of United Home Care, Inc., Fairfield, CT.

Anxiety	Borderline P.D.	Chronic Pain
Crisis Intervention	Dementia	Depression
Generic	Homicidality/Suicidality	Mania
Noncompliance	Nutritional	Phobia
Psychotic Symptoms	Relapse Prevention	Restraint Use
Safety	E.C.T.	

FIGURE 7.3   Copathways.

A Patient Goals Worksheet is filled in by the nurse in collaboration with the patient listing goals to be achieved in one week, two weeks, one month, and two months. This helps with the overall plan for care so that both nurse and patient have a sense of direction and expectations for care. Clarification can occur at this point so that all parties involved in care including the caretaker have realistic goals about what can be achieved. The patient is asked to sign the Patient Goals Sheet and a copy of the weekly plan is given to the patient to use as a worksheet to track progress over the course of the week.

Quantifying Behavioral Outcomes

Quantifying outcomes in the area of human behavior has assumed priority in the era of managed care. However, it is difficult to assess and attribute outcomes to specific home care treatments because each intervention is only one of a long continuum of services that the patient receives. Isolating the one variable describing the impact of services is most likely not statistically feasible. Patients often have multiple problems ministered by multiple providers. But information can be provided such as what type of treatment on average has been found to most benefit a patient with a particular disorder or problem. Besides assisting in clinical decisions, outcome measures also help achieve regulatory compliance as well as providing marketing information regarding cost-effectiveness and treatment-effectiveness of services to clients, other providers, and managed care.

Outcomes measured generally fall into two general categories, patient outcomes and cost-effectiveness outcomes. Cost-effectiveness outcomes include length of time on-service for home care services, number of visits by home care nurses, number of doctor visits, number of rehospitalizations, usually over a 12-months time frame, and psychiatric nurse costs.

Patient outcomes include patient satisfaction, patient functioning, such as ADL and patient symptoms. Patient outcomes for all patients who are in the psychiatric program at the agency include the Patient Satisfaction Survey, The Global Assessment of Functioning Scale, the Patient Status Rating Scale, and specific clinical outcome measures for the problem being addressed. Patient outcome measures are given at intake, halfway through the critical path during the treatment phase, at discharge, and ideally again about 6 months after discharge from service.

Patient satisfaction is measured three months after the patient is discharged through a random telephone survey. Research has demonstrated that telephone surveys do not compromise the validity or reliability of an instrument (Dockerty & Dewan, 1995). Telephoning patients also controls for response bias in that patients may feel they need to rate their care positively if given a questionnaire by their primary nurse at discharge.

Standardized forms are used whenever possible in order to ensure the reliability and validity of the measures and so that there is a common language for communication with other health providers and managed care. Also the use of tools which are widely-used permits comparison of aggregate data from the agency with other large databases.

Self-report forms, interview formats, and objective rating scales are all utilized as instruments for assessment and outcome data. Although self-report forms are simpler to administer and require less staff time, many home care patients are elderly and need assistance in answering questions on these forms. Objective rating scales for the nurse to fill out are also used.

Care should also be taken to check with the authors of the instruments you want to use in order to ensure that the tool is reliable and valid for the elderly, since normative data for instruments is frequently on a college age sample. An alternative form may be available that was developed specifically for your population. If not, two different instruments may need to be utilized, depending on the age of the patient. For example, The Beck Depression Inventory and The Geriatric Depression Scale are utilized as two measures for the Depression Critical Pathway. However, parsimonious use of the minimum number of instruments is encouraged so that meaningful data can be compared and contrasted. If 10 different measures for depression are used it is not possible to compare outcome results of so many instruments except through sophisticated meta analysis procedures.

As the patient progresses toward the desired outcomes, the critical pathway tracks how patient goals are being met. Thus quality assurance is continuously monitored. With computerized charting, it is possible to tabulate outcome data as it is being documented which allows for better communication between caregiver, management, insurance companies, and accreditation agencies. The outcome monitor is evaluated as part of the agency's quality assurance program evaluation activities. This process allows for ongoing, continuous research to improve patient care.

THE PSYCHIATRIC HOME CARE TEAM

The psychiatric nurse serves as liaison between others on the home care team and the patient and works to build a relationship with the patient which serves as a bridge to the community. The critical pathway serves as a communication tool to coordinate these services. Interteam services such as home health aide services and community referrals are addressed as care elements for every visit on the critical pathway.

The psychiatric home care team consists of the psychiatric nurse who is the case manager, the social worker, the psychiatric home health aide, and the physician or psychiatrist. As of 1996 Medicare requirements changed so that the patient's primary care physician can now order psychiatric care. Prior to this, only a psychiatrist could order psychiatric home care services. This has resulted in easier access of patients to our program because patients, particularly the elderly who are often not receptive to seeing a psychiatrist, usually do allow the psychiatric nurse to visit. A psychiatrist is employed by the agency as a consultant who reviews cases with the nurses once a month, makes home visits when necessary, and is available for telephone consultation.

Physicians' orders are usually general such as skilled nursing to educate regarding medications, check medication compliance, or teach about disease processes. It is left to the psychiatric nurse as to how that care will be provided. Nurses are formally accountable for the outcomes specified on the tool (Zander 1991). With a critical pathway framework, nurses have a comprehensive plan of care which is structured and systematic.

Because of the high degree of autonomy regarding treatment decisions and the wide range of psychotherapeutic skills needed, nurses in the Psychiatric Program are required to have at least two years of recent inpatient psychiatric experience as well as graduate psychiatric nursing education. Clinical Nurse Specialists with Masters in Psychiatric Nursing ensure clinical excellence and expertise. Ideally the psychiatric nursing team's expertise should reflect specialties of the patient population it is serving. For example, the Home Detoxification Program is managed by a Clinical Specialist who specializes in substance abuse. The Anxiety Assessment and Treatment Program is managed by a nurse who specializes in phobia desensitization. The nurse clinician's skilled judgment serves to evaluate the critical pathway and revise it as appropriate. The critical pathway then becomes an excellent clinical resource setting practice standards. It also serves as an orientation tool for new nursing personnel as job expectations, scope of practice, and patient care are presented in a structured, systematic manner (Zander, 1987).

Social work services in the agency include counseling for long-range planning and emotional factors relating to financial planning, housing, legal issues, health insurance, community resources, and short-term goal-oriented intervention therapy. Since the psychiatric nurse and the social worker frequently collaborate on cases and both provide short-term psychotherapy, clear documentation of goals by each discipline is imperative to avoid redundancy of services. For example, the patient may be on psychiatric medication and the psychiatric nurse's care focus may be on assessment of compliance with medications or side effects while the social worker may focus on working with family relationship issues. This can be stated on the critical pathway so that documentation is clear and all care providers can easily see at a glance who is doing what and when.

CONCLUSION

The development of critical pathways continues to be an ongoing process as clinical data dictate changes. Clinical conferences are an important forum for discussing ways to simplify, clarify, or revise existing pathways. Care plans which were often experienced as drudgery have been replaced with a tool which enables the nurses themselves to continuously challenge and change their practices as the care for their patients dictates. This encourages creativity and involvement which helps with job satisfaction and ultimately results in better care for patients.

As restructuring of the health care system progresses, it is clear that accountability and outcome-oriented care will continue to be priorities for the future. Use of critical pathways in psychiatric home care holds the potential for decreasing recidivism and improving health outcomes in less time, thus saving further visits to health care providers. As critical pathways are further refined and research documents results of this model of care, quality care should become a reality for all people needing home care services.

REFERENCES

American Nurses Association (1994). A Statement on Psychiatric-Mental Health Clinical Nursing Practice and Standards of Psychiatric-Mental Health Nursing Practice, Washington D.C.: ANA Publishing.

Cooke, M.K., & Brodrick, T.M. (1994). *Critical Pathways*. In Marilyn Harris, (Ed.), *Handbook of Home Health Care Administration* (pp. 309–319). Gaithersburg, MD: Aspen Pub.

Dee-Kelly, P.A., Heller, S., & Sibley, M. (1994). Managed Care: An Opportunity for Home Care Agencies. *Nursing Clinics of North America, 29*(3), 471–481.

Dockerty, J.P., & Dewan, N.A. (1995, October 24). Guide to Outcomes Management, National Association of Psychiatric Health Systems, Presentation at Greater Bridgeport Mental Health Center in Bridgeport, CT.

Dunn, J., Rodriquez, D., & Novak, J.J. (1994). Promoting Quality Mental Health Care Delivery With Critical Path Care Plans. *Journal of Psychosocial Nursing, 32*(7), 25–29.

Gingerich, B. & Ondeck, D. (1996). *Clinical Pathways for the Multidisciplinary Home Care Team*. Gaithersburg, MD: Aspen Pub.

Homan, C. (1994). Critical Path Network. *Hospital Case Management, 10,* 135–138.

VNA First (1994). *Home Care Steps Protocols and Home Costeps: Pathways to Health*. Berwyn, Il.: VNA First.

Zander, K (1987). Critical Paths: Marking the Course. *Definition, 2*(3), 1–4

Zander, K. (1991). What's new in managed care and case management. *New Definition, 6*(2), 1–2.

BIBLIOGRAPHY

Townsend, M. (1994). *Nursing Diagnoses in Psychiatric Nursing,* 3rd ed., Philadelphia: F.A. Davis Co.

Zander, K. (1998). Nursing Case Management: Strategic Management of Cost and Quality Outcomes. *Journal of Nursing Administration, 18*(5), 23–30.

8

Critical Pathways in Rehabilitation and Long-Term Care

Jane Woolley

The health care team wishing to initiate critical pathways in a long-term or rehabilitation setting is presented with special challenges because of features
unique to these settings. Critical pathways are thought to be difficult to use in rehabilitation and long-term care because of the many regulatory agencies (CARF,
JCAHO, OBRA) and the complex needs of the older individual. However, a well-
focused critical path can improve the quality of life and outcomes of patients or
residents with chronic illness or a long rehabilitation.

It should be clear to the reader who has perused this book in an orderly fashion that while critical pathways are useful, time saving, cost effective, and even
necessary in today's managed care environment, pathways cannot be created by
a single practitioner or a single discipline and be effective. This chapter provides
an overview of the process of developing critical pathways for patients or residents in these settings.

PERSPECTIVES IN LONG-TERM CARE

Literature concerning critical pathways in long-term care is almost nonexistent. In
a recent informal poll of 20 nursing home educators, no one reported the use of
critical paths in their facility. The pathways that do exist for these settings are often
developed around diagnostic-related groups (DRGs) just as they have been in acute
care and rehabilitation. Often practice guidelines have not been developed for many

of the chronic illnesses found in these settings such as chronic obstructive lung disease, hypertension, and diabetes.

Critical pathways have worked well for episodic illness treated in the hospital, but for chronic, long-term patients, the focus shifts from cure of illness to assisting the individual to adjust to lifestyle changes. The problem of using a medical diagnosis focus to manage the care of a resident in a long-term care setting is that the individual is relegated to the role of dependent patient as a lifestyle. This dependence, inherent in the traditional patient role, robs the resident of his or her autonomy, ability, and, finally, his or her will to live.

Another difficulty with a medical diagnosis approach is that many of the diseases in a long-term care facility are chronic in nature and exist with other chronic diseases. This chronicity has not been well managed over a lifetime, and in some instances, new symptoms are overlooked and thought to be just part of the chronic illness. The symptoms for the older patient who is admitted to a long-term care facility from a hospital often are compounded by the deleterious effects of bedrest, medications, and a misunderstanding of the physiological and other psychosocial needs of older patients.

A nursing diagnosis approach is another perspective that has been adopted for critical pathways in long-term care, for example, anxiety, pressure ulcer, and incontinence. However, a nursing diagnosis also reinforces the medicalization of aging because it is a problem-solving process, and, as such, a problem is identified that needs to be fixed and this precludes treating the individual holistically in a wellness-oriented framework. The emphasis for patients in long-term settings should be wellness, and those functions that we know are essential to an individual's sense of well-being should be the foci of concern.

Ferri (1994) developed several categories based on nursing diagnoses and Henderson's basic functions that might prove useful in the development of pathways that are supportive of the aging process. These categories include health perception, nutrition, elimination, activity/exercise, sleep/rest, role relationship, cognitive/perceptual, self-perception, sexual/reproductive, value/belief, and coping/stress. These are not significantly different from the Main Data Set (MDS) assessments and Resident Assessment Protocols (RAPs) but are more functional than the labels of falls, feeding tubes, pressure ulcers seen on the RAPs, or those of nursing diagnoses that focus on the disability of the individual. A critical pathway created with these functions in mind could be structured to be consistent with the time frames for the resident assessment process. (See Figure 8.1).

For example, on the day of admission, the resident or patient would be assessed for immediate needs and concerns related to his or her diagnosis and his or her functional state (wellness/health perception) and a determination made as to whether the interventions are focused on the restorative, rehabilitative, or maintenance mode. Days 1 through 10 would then be focused on gathering information on the individual's wellness needs and his or her nutritional needs (albumin, weight, intake and output), as well as the other categories. While these labels are different from those used in rehabilitation or acute care, which uses consults, tests, activity, and medications, the use of functional labels could be integrated with those that might be useful, such as medications and treatments. Identification of the need for medication holidays as well as required treatments should be included.

In using a functional perspective for critical pathways, the overall goal is to restore or maintain the functioning of the individual to ensure that residents achieve the highest level of functioning possible and maintain their sense of individuality (Morris, Murphy, & Nonemaker, 1995). An integrated critical pathway would then look at the person's functional status (activities of daily living) as well as the support structures (medications, meals) and the preventive interventions (vaccines, dental care, mammography). This provides the needed structure for maintaining the autonomous functioning of the older individual in a system that is often responsible for his or her increased dependence. The quality monitor of the critical path could then show at a glance the resident's decline or improvement and encourage the team to discover the cause and change the interventions to reverse the decline or reinforce the interventions to maximize the improvement.

LAYING THE GROUNDWORK

Critical pathways have four major functions: (1) patient or resident outcomes; (2) established time lines; (3) collaboration; and (4) comprehensive aspects of care (Ignatavicius & Hausman, 1995). They assist the health care team, the patient, and, increasingly, the payor in identifying appropriate outcome goals. Critical pathways can and should coordinate the care provided to the patient or resident so that he or she gets the right thing, at the right time, in the right amount. Critical pathways provide a mechanism for identifying what interventions impact patient or resident care and can help identify what resources are being utilized.

An initial consideration in implementing pathways is to identify the anticipated outcome or goal. This is no small enterprise. While there are specific goals for each of the pathways, developing and implementing pathways requires the commitment of the organization in order to succeed, and the first step of the administrative team is to identify the overall goal for implementing pathways.

Along with goal identification, it is important to understand what it can do and what it cannot do or you will be defeated before you begin. Just as rehabilitation is not reincarnation, critical paths are not the answer to all the problems in health care. Both the literature and the grapevine are replete with stories of pathways that did not provide the anticipated benefits. Indeed, when I was introduced to pathways in a rehabilitative setting, they were a dismal failure. In retrospect, I believe that failure was related to how the pathways were developed and introduced.

The pathways were developed outside of the multidisciplinary team responsible for the care of the patients, and the sole reason presented for developing the pathways was to decrease the patients' length of stay. While critical pathways have

CHRONIC DIAGNOSES: Patient Name: Age:

Pattern	Day of Admission (0 day)	Day 1-10	Day 11-20	Day 21-30	Day 30-90	Day 91-180	Day 181-271	Annual Day 364
Health Perception/ Wellness	□ Nursing Assessment □ Family Contact	□ H&P				□ Consider lower level of care		□ H&P
Nutritional	Assess: □ I & O □ Weight □ Dietary Needs/Preferences □ Chewing/Swallowing	□ Weight □ I&O	□ Weight □ D/C I&O Diet as Tol	□ Monthly Weights Stable ___	□ Alb wnl			□ Weight ___
Elimination	□ History Established Voids Q ___ Last BM ___	□ BM qd	□ Continence Program □ Bowel Routine ___					
Sleep	Sleep Pattern Assessment includ: □ # Hours Awake ___ □ # Hours Asleep ___ □ Time of Rising ___ □ Naps ___ □ Nightmares ___ □ Uses Own Bed ___	□ No covers on feet	□ Reestablish sleep pattern	□ Guided Imagery □ Warm Milk at 9 pm	□ Sleeps ___ hours without interruption ___			
Cognitive/ Knowledge	□ Assessment □ MMSE □ GDS	□ Complete Assessment		□ Current Events Group □ Aging Issues Group	Group Partic: □ Daily □ Intermittent □ None			
Activity/Exercise	Safety Assessment including: □ Safe to Sit □ Safe to Stand □ Safe to Ambulate □ Ambulation devices required □ PT/OT/ST Consult prn			□ Exercise Group M-F	Group Partic: □ Daily □ Intermittent □ None	□ TaiChi MWF		
Self-Perception	Establish Preferences, Patterns □ Self-Care Assessment □ Grooming Patterns							
Coping	□ Social Service Assessment □ Coping Skills Assessment	□ Utilizes coping skills		□ Participates in diversional activity ___	□ Uses New Coping Skills ___			

FIGURE 8.1 Long-term care critical pathway.

Pattern	Day of Admission (0 day)	Day 1-10	Day 11-20	Day 21-30	Day 30-90	Day 91-180	Day 181-271	Annual Day 364
Value/Belief	□ Assess Value/Belief Needs □ Spiritual Assessment	□ Spiritual Services						→
Tests	Lab Tests: □ Hct □ Hgb □ RBC □ UA □ TB □ BS □ Bun □ Creat □ VDRL □ Alb □ Vital Signs QD	□ As Indicated				□ Life Review		□ Rectal □ Occult Blood □ H,A, RBC
Medications	□ Assess Effects/Side-Effects					□ Medication Holiday		□ Flu Vac
Teaching	Assessment of: □ Life-Style □ Use of Assistive Devices □ Medication Usage/ Understanding □ Complications From Immobility			□ Teaching Plan Established □ Demonstrates Knowledge of Medications	□ Teach Stress Relief Activities			
Treatments	□ Advance Directives, Code Status, Client Goals □ Assess for Anxiety/Depression □ Ambulate to BR and Meals	□ Set-up Treatment Plan with Client/ family □ Ambulate 100-150' tid		□ Ambulate 300' qid				
Sexuality	Assess: □ SBE □ Sexual History/Concerns							
Role Relationship	□ Assess Current Roles □ Assess Family System: Issues & Goals □ Impact of client condition on Family	□ Family Meetings prn			□ Relationship with 1-2 people			□ Verbalizes satisfaction with current relationships

FIGURE 8.1 (continued).

113

repeatedly been shown to decrease length of stay (Lumsdon & Hagland, 1993; Gates, 1995), most practitioners involved in patient care have been taught to provide quality care to their patient, and they are unlikely to change their behaviors unless they know that critical pathways will improve patient outcomes. Caregivers are generally suspicious if the administration department wants to decrease the length of stay for the benefit of the hospital or agency, not the patient.

While length of stay is a consideration in some settings, it is not a useful measure in the long-term care facility, except as it relates to the person's longevity. In the long-term care facility, the focus is on the health outcome, the functioning of the individual, rather than on length of stay or the cure of a specific disease entity. In addition, the length of stay in a long-term care facility or a rehabilitation hospital may be variable, depending on a wide range of factors. These factors need to be accounted for when developing the critical pathway. For example, patients admitted for rehabilitation under DRG 14 (cerebral vascular accident, CVA) have a large disparity in brain injury and thus physical disability. Patients with minimal disability might be ready to be discharged after a brief period of rehabilitation. Patients with moderate disability require a slightly longer rehabilitation, and patients with severe brain injury and disability often require a long period of rehabilitation. A CVA critical pathway needs to be designed to meet the needs of this distinct group of patients. One way of addressing different patient scenarios in one critical pathway is to build decision points into the critical pathway. Considering patient outcomes at these decision points allows the staff to work with patients at a pace that is individualized to the patient (Hydo, 1995).

Administration and Physician Support

The administration department needs to understand the value of critical pathways and to determine who the principal players are and empower them to implement the project. In both the long-term care facility and the rehabilitation hospital, as in the hospital, the administrator has the ability and the responsibility to ensure that the physicians and department heads participate in the process of critical path development, implementation, and variance management. Since physicians are responsible for admitting the patients and writing many of the orders that impact quality and cost issues, they must be part of the initial discussions. Once you have convinced administrators and the physicians, the rest of the team is much easier to engage. Convincing these two groups is a matter of doing your homework.

Both administrators and physicians are attentive to documented facts. Gather up the data that demonstrate that critical paths make a difference in patient outcomes, that is, that rehabilitation patients go home more functional, have higher FIM (Functional Independence Measures) scores, cost less in ancillary services, are more satisfied, and have a shorter length of stay, and present the data in a manner that minimizes reading. In the long-term care facility, look at the functional abilities of your residents. How many significant changes do you see on the MDS? If you are in a multilevel facility, how long do they remain independent? How many become more independent after they are admitted to the facility? What is the incidence of pneumonia, pressure ulcers, or falls? Use graphs: pies, bars, and lines to show the difference in your population and the population served by critical pathways. If there are not data available, use the national, state, or local data.

Create a tool listing the variables you want to target with your path and collect that data. Chart review is a key in establishing the amount of variance in treating a particular population in your institution. If you don't know your current practices, you won't be successful at fixing the problem.

In collecting the data, ensure that they remain confidential and that they are accurate and comprehensive. If the data are questionable, the whole project may be sabotaged. When presenting the data, be sure you have several people in the room who you have already convinced so that if there is some controversy, you aren't alone in presenting the data.

We recently collected some data on one of the paths that has been in the facility for about a year. Unfortunately, while we had discussed our game plan, we neglected to say we wouldn't present the data unless we were all there. One of the more inexperienced managers ended up alone with the physician and the nursing administrator at a meeting, shared the data, and was attacked from both sides. At the next meeting, we made sure that the team in charge of data collection was together and strategically placed around the table. We also spent some time with the administrator to explain the data. People don't like surprises. If the administration and the director of medical services don't understand and support the concept of critical paths, don't start.

Choose a Champion

While it's been mentioned before, it bears repeating. Prior to instituting a critical pathway, certain conditions must be met, and one of the most important is a champion for the cause. Initially, the champion should be a physician. This makes the task of convincing the other physicians to follow the path much easier.

How do you identify a champion? He or she should be someone who has a keen interest in making that path a success, someone who works well with others, and who has the power of persuasion. I was involved in the development of one pathway where the physician was both interested and knowledgeable. Unfortunately, he was not given to talking, either to his peers or in large groups. The path languished.

In rehabilitation, the phychiatrist might be the logical choice, or, depending on the diagnosis, it might be a physician who has a specialty practice or a general practice. This depends somewhat on how the unit is established and the physicians in the community. Is the phychiatrist the physician for all the patients, or a consultant? If he or she is a consultant, and the pathway is CVA, he or she might be the logical person, or perhaps the neurologist. In the long-term care facility, the physician champion might be one of the physicians on the staff or one of the community physicians who has patients in the facility and the respect of his or her peers. In either case, the physician must be introduced to the concept, the rationale, and the supporting data.

Physicians in long-term and rehabilitation settings, like their counterparts in the hospital, are often reluctant at first to participate in critical pathway development. They often see pathways as "cookbook medicine," too time consuming, creating opportunities for malpractice cases, and as someone telling them how to practice medicine. The key to their participation lies in convincing them that the process is in the best interest of the patient (better outcomes), the physician (fewer phone calls to initiate specific orders), and the hospital (decreased length of stay and better utilization of resources).

In my experience, the person best suited to convince the physician is the nurse with the knowledge both of the practices within the facility and the standards in other facilities. In the rehabilitation setting, this person may be the clinical specialist or the nurse manager, while in the long-term care setting, it might be either of those or the case manager or director. The title is not as important as knowing the data and the process.

In any case, the champion should know current outcomes for the particular patient population, the length of stay, the cost per case, the number of X rays, and so on. Each path may need a different champion, particularly since this person will spend a lot of time in group meetings and one person may not be able to give the amount of time this process requires. We have found that having the chief of medicine appoint a particular physician is useful but only if that physician has the right skills.

Involving the Team

In my last position when we introduced the critical pathway to the rehabilitation team, there was great concern about the inability to predict when the patient would be able to do a specific task or reach a certain goal. The unit manager and the clinical specialist had met with the team and had explained the purpose of the pathway and the data but were not able to get the team to agree on goals. This in spite of the fact that the team was used to meeting weekly to identify specific goals for each patient with an anticipated discharge date. I was asked to come and meet with the team, in my role of pathway expert, to assist them to understand the relationship among their standards and protocols and the steps on the pathway. I went to the meeting armed with information from the state, the rehabilitation association, the protocols from rehabilitation services, and the unit's statistics. By showing the team how others were doing compared to us, and by convincing them that this was no different than their existing protocols and standards, and by showing them the variance in their practices, they were able to move forward. That pathway has been in existence for over a year and has assisted the team to meet their goals of greater independence and shorter length of stay. But lest I mislead the reader, it took another 2 months for the team to actually agree on when interventions should be introduced and when the patient should reach specific goals.

Each discipline and department that interacts with the patient should be involved in the initial discussions. Then if they determine that they can't contribute to the pathway, they have made that decision and won't be an impediment to your progress because they feel left out. Interdisciplinary communication is one of the greatest benefits of critical paths. A recent study found that, generally, physicians, nurses, and other staff have little knowledge of the other's clinical practice (Coffey et al., 1992). The opening of communication through a dialogue about critical pathways helps decrease variation related to systems, habits, and practices that have evolved without the benefit of research or sound fiscal practices.

Be sure to include the finance people, coders, and social services as well as physicians, nurses, and administration. If consultants were used in a previous attempt and they are still available, include them in the process. Focus on the process, not the caregivers and their practices. Set specific goals. Without a goal, it is unlikely that you will be successful. If the population doesn't have a lot of variability that can be standardized, don't create a path. In other words, whatever you do should result in significant gains, otherwise it probably is not worth the amount of work that goes into creating and maintaining paths. In our rehabilitation unit, we chose the more common diagnoses that represent a substantial part of the caseload, for example, cerebrovascular accident, total hip replacement, and chronic obstructive pulmonary disease.

Monitoring Outcomes

The major function of critical pathways is to provide quality outcomes that are measurable, increase utilization of the unit, minimize variation in the treatment of patients, provide services in a more timely manner, and increase the market share in the community. Thus since identifying, tracking, and carefully monitoring outcomes is the central concern for a quality critical path, make sure you have established a monitoring system in place *before* you implement the path.

A clue as to how successful any path is going to be is clarity of outcome measures in the quality outcome monitor. If outcomes are not clearly delineated, then it will be difficult to monitor whether our interventions are successful and better care has been delivered.

For example, consider the following. A pathway has been written for prevention of urinary tract infection. One of the interventions identified under the diet label is to "increase consumption of cranberry juice." This poses several problems. The word "increase" assumes the patient is already ingesting some of this juice. If the patient drinks 50 milliliters of cranberry juice, will it be sufficient? Most practitioners have agreed for some time that at least a thousand milliliters of cranberry juice is necessary to make an appreciable difference in the acidity of the urine. A better goal or outcome would be that the patient or resident drinks a specified amount of this liquid. However, given the age of the patients in long-term care, it might be more to the point to provide sufficient ascorbic acid to render the urine acidic without having to drink a quart of cranberry juice. But, if we put aside the issue of amount and focus on how we will measure the outcome, it becomes apparent that we need more detail.

Another example of lack of clarity about outcomes is an early critical pathway outcome to "avoid dehydration." Here again there are no measures for the quality monitors making it difficult to assess if the patient did not follow the path due to an individual illness pattern or to a lack of coordination of fluids, or a system problem, such as a water main break.

Having the quality outcome monitor at the bedside, or with the charts, helps to minimize the amount of work to do once the patient is discharged. In long-term care, frequently the pharmacist can provide some of the information on medication usage since this is something he or she gathers for other reports. Always identify outcomes rather than processes. For example, instead of talking about assessment as a process, we identify when the assessment is complete.

While the critical pathway should not repeat everything that is in the standard of care or on the protocols or practice guidelines, it should provide sufficient substance to measure and should clearly contribute in an appreciable way in a specified time frame to the desired outcome of the patient.

In rehabilitation there are so many disciplines involved and because the goals are often shared by disciplines, it makes developing the path and the monitor a challenge. By making the measurable outcome the goal, any of the disciplines can identify its achievement. Also, because the team is used to meeting weekly or biweekly to discuss the progress of the patient and reconfirm or reformulate goals, these provide meetings an ideal forum to discuss the pathway variances. Reports can be built on the path and this should decrease the amount of time spent in team conferences because everyone knows the goals. While the team can and should report on the week-to-week progress they are making on the established outcomes, the group should also report their collected variances and action plans to the larger administrative group.

The pathway should be able to be used with 80% to 85% of the population or else all you do is track variances (VHA, 1995). Variability is easier to track once critical pathways are in place, and tracking variances is important because variability has been implicated in decreased quality outcomes and increased costs (Musfeldt & Hart, 1993; Schriefer, 1995).

The following example illustrates use of variance tracking resulting in saving money and better patient care. In one unit, the patients who had total hip replacement were being discharged to a subacute unit in a local nursing home facility despite the fact that the hospital had a rehabilitation unit and a critical pathway that demonstrated better outcomes than the nursing home unit. Investigation revealed that the physicians felt their patients waited too long for a bed on the rehabilitation unit and they were being pressured by administration to get the patients out as soon as possible. The rehabilitation unit management came up with an action plan to screen the patients on the hospital unit on the 4th postoperative day and admit them to the rehabilitation unit on the 5th postoperative day. They also showed the orthopedic physicians their critical pathway data demonstrating that the patients were sent home from rehabilitation on the 6th day and that they were able to be independent in their activities of daily living. Subsequently, the physicians admitted their patients to the hospital-based rehabilitation unit and everyone benefited. The physicians could continue to follow their patients, the patients didn't need to go to a nursing home, and the hospital was able to benefit from both DRGs while increasing their market share. In fact, the length of stay decreased by almost 2 days with the institution of the path; the patients achieved higher scores on the Functional Independence Measurement; and the unit increased its revenues going from a loss to a net gain in revenue.

Delineating Time Frames

Time frames for outcomes must be delineated by the team. Time frames are different for rehabilitation and long-term care from acute care. In long-term care, the time line is extended even farther. Pathways are planned for weeks, months, and years instead of for hours and days (VHA, 1994; Ignatavicius & Hausman, 1995).

Consider the cerebrovascular accident (CVA) patient in the emergency room setting where time lines are minutes or hours. In this version of the pathway, some interventions must take place immediately (e.g., airway establishment and assessment of neurological status), while other activities wait until the patient is stabilized and admitted to a nursing unit (nutrition and bowel regimens). For the CVA patient on the rehabilitation unit, different time frames are operative. The patient may be completely assessed with goals established by day 3. By the end of the first week, the goal is for the patient to be independent in wheelchair activities, and by the time of discharge, at the end of the second week, the goal is to be able to transfer from the wheelchair to an automobile. In long-term care, the time line is extended even farther and may focus on becoming increasingly independent or mobile.

The time frames for critical pathways in long-term care can be established as an illness resolving or based on the established time lines inherent in the minimum data set process required by OBRA (Omnibus Budget Reconciliation Act). In the former instance, the time frames might be in phases, for example, assessment, stabilization, rehabilitation, palliative, and discharge. In the latter instance, the time lines would be 14 days (data collection), 21 days (analysis of data and care planning), 3 months, 6 months, 9 months, and 12 months (quarterly and annual reviews). Additionally, anytime there was a significant change in the patient's status, one would need to reassess the path to ensure that the patient did not need alternative interventions.

Keeping the Process Moving

Despite the success of the rehabilitation unit staff with the total hip replacement path, they have been unable to initiate a cerebrovascular accident critical pathway that they developed months earlier. The administration department of the institution supports critical paths, requiring each service line to develop and report on the results of the paths to the hospital quality steering committee. While everyone on the team supports the path and the rehabilitation team knows the value of the pathway, the pathway remains unimplemented.

Why? There have been major changes in personnel, leaving a critical lack of an individual to manage the day-to-day progress of the path and the outcomes achieved. The lack of a specific individual is more often a problem when the process is new but applies to any situation. You need to have the appropriate people keeping the process going.

One of the strategies that we have found helpful in keeping the process going and on track is to provide the patient with a copy of the pathway that is written in a manner he or she they can understand and follow. This engages the patient in his or her treatment plan and also provides someone who is really interested in the outcomes to keep an eye on the process. We have had many patients ask why we are deviating from the path, which helps us get back onto it. But, while this helps, it is critical to have one person responsible for collecting the quality outcome indicators. If people are not held accountable, no matter how good their intentions, the path will have no impact on the patient outcomes.

Barriers to Change

One would think it should be relatively straightforward to introduce critical paths to both rehabilitation and long-term care because the populations are relatively stable, and in the latter instance, fairly predictable. If the population or pathway is chosen carefully, the major impediments are those that one encounters whenever one introduces change, that is, system barriers and individual egos. One of the interesting things I have found in long-term care is that the problems are not significantly different from acute care.

One of the major system barriers in both rehabilitation and long-term care is the lack of computerization. If hospitals have been slow to join the computer age, long-term care in most instances shares this lack of alacrity. In the absence of computerization, the current charting mechanisms may need to be replaced, and this should add minimally to the workload of personnel if you expect to get the data.

Lack of secretarial support is the second major barrier to successful implementation. Most of our systems were developed when the amount of paperwork was significantly less than it is today. The collection of variance data, which is the cornerstone to successful outcomes, is often dependent on nursing personnel. As with most documentation, it is often the last factor to be considered in the pressure of completing patient care.

In institutions that do charting by exception, the critical path as a document can enhance the system. Most long-term care facilities do not chart by exception, and few rehabilitation centers do. What we have found to be effective is a format that promotes charting on the path itself. This can be accomplished by incorporating boxes for initials or simply by having the appropriate discipline initial the action. This is most helpful when the outcomes are identified on the path rather than the processes.

The computerization of the assessment process in long-term care and rehabilitation settings will invite closer scrutiny of long-term care, providing easy access to both outcomes and interventions, making critical pathways a necessity to both assist in the ever-difficult task of communication among the team members and providing quantitative and qualitative information about the outcomes of long-term care.

CRITICAL PATHWAYS AND THE FUTURE

The future of critical paths seems assured given the current climate of reducing health care costs and improving profit margins for both managed care and insurance companies. It took more than 10 years for critical pathways to be accepted in the acute care environment. The growth in long-term care should be much quicker given the different circumstances that we face in the late 1990s.

The health care continuum is shifting to a wellness-focused system, and the acute care facility is being utilized for only the sickest of patients. The community and long-term care are expected to provide care to patients who are sicker than they have been at any other time. Further, those physicians who have learned about critical paths in the acute care setting will be looking for the same benefits of these communication tools in long-term care. Patients or residents will also begin to look for more predictable outcomes, whether that means returning to the community, maintaining their function, or dying in an environment that provides humane care.

Competition will continue to become more fierce with the formation of health care alliances that are charged with the responsibility of making the community healthier. Facilities that are unable to provide evidence of their outcomes, whether long-term care, rehabilitation, assisted living, or community agencies, will become extinct.

CONCLUSION

In summary, the question is not, do critical pathways belong in long-term care and rehabilitation, but what model should be utilized to promote them? The challenge is to have the multidisciplinary team provide coordinated, timely, outcome-focused care to residents and patients utilizing only the necessary resources. The development of critical pathways must be interdisciplinary and must be planned for. The successful team will anticipate resistance and focus on data, not individuals, as they move forward to meet the future.

REFERENCES

Coffey, R., Richard, J., Remmert, C., LeRoy, S., Schoville, R., & Baldwin, P. (1992). An introduction to critical paths. *Quality Management in Health Care, 1*(1), 45–54.

Ferri, R. S. (1994). *Care planning for the older adult. Nursing diagnosis in long-term care.* Philadelphia: Saunders.

Gates, P. (1995). Think globally, act locally: An approach to implementation of clinical practice guidelines. *Journal on Quality Improvement, 21*(2), 71–84.

Hydo, B. (1995). Designing an effective clinical pathway for stroke. *American Journal of Nursing, 95*(3), 44–51.

Ignatavicius, D., & Hausman, K. (1995) *Clinical pathways for collaborative practice.* Philadelphia: Saunders.

Lumsdon, K., & Hagland, M. (1993). Mapping care. *Hospital and Health Networks, 67*(10), 34–40.

Morris, J., Murphy, K., & Nonemaker, S. (1995). *Long-term care facility resident assessment instrument (RAI) user's manual.* Baltimore: Health Care Financing Administration.

Musfeldt, L. C., & Hart, R. (1993). Physician directed diagnostic and therapeutic plans: A quality cure for America's health crisis. *Journal of the Society for Health Systems, 4*(1), 80–88.

Schriefer, J. (1995). Managing critical pathway variances. *Quality Management in Healthcare, 3*(2), 30–42.

BIBLIOGRAPHY

American Thoracic Society (1993). Guidelines for the initial management of adults with community acquired pneumonia: Diagnosis, assessment of severity, and initial antimicrobial therapy. *American Review of Respiratory Disease, 148,* 1418–1426.

DeWoody, S., & Price, J. (1994). A systems approach to multidimensional critical paths. *Nursing Management, 25*(11), 47–51.

Eliopoulous, C. (1995). *Care of the elderly in diverse settings.* Philadelphia: Lippincott.

Gallagher, C. (1995). Pediatric clinical path program development: Project selection and rollout. *Journal of Health Care Quarterly, 17*(3), 4–10, 16.

Lumsdon, K. (1993). Rule 1 on critical paths: Proceed with caution. *Hospital Health Network, 67*(22), 56.

VHA, Inc. (1994). *Critical Pathways in Case Management Sourcebook.* Irving, TX: Author.

9

Liability Issues in Development, Implementation, and Documentation of Critical Pathways

Joanne Sheehan
Gayle H. Sullivan

INTRODUCTION

When carefully researched, well written, properly followed, and completely documented, critical pathways are a valuable tool for minimizing liability. The critical pathway, however, like all systems of care planning and documentation, carries attendant liability issues. Significant concerns include:

- whether potential liability exists for the critical pathway developers;
- whether potential liability exists for those who follow the critical pathway in planning and providing care;
- whether documentation recorded on a critical pathway is sufficient to demonstrate that the standard of care was met;
- whether critical pathways, once implemented, create standards of care and can effectively substitute for professional clinical judgment.

As we evaluate critical pathways from a legal and risk-management perspective, bear in mind that exercising reasonable professional judgment and documenting that you did so are a key to protecting yourself from liability.

WHAT IS MALPRACTICE?

Malpractice is defined by four distinct elements, all of which must be proven by the plaintiff.

Elements of Malpractice

- *Duty:* Established by a professional relationship
- *Breach of Duty:* An act or omission in violation of the standard of care
- *Injury:* A physical injury such as fracture, paralysis, or brain damage
- *Causation:* Breach of duty caused the client's injury

A duty is created by a professional relationship. With that in mind, consider the following example:

A home health nurse stops at a pharmacy on her way to work. She sees that an elderly woman is lying on the sidewalk just in front of the store and a small crowd has gathered. A former client recognizes the nurse and calls out for her to help. The nurse does not want to become involved and she shouts back, "I'm not a nurse anymore," and then enters the pharmacy. The elderly woman dies.

Will the family of the deceased be successful in an action against the nurse if they claim that she had an obligation to stop and help and her failure to do so caused the woman's death? Readers' conflicting answers to this question illustrate a difference between what the law and ethical codes require. Here, while ethical codes dictate that the nurse should assist, the law does not require it. A malpractice claim against the nurse under these circumstances will surely fail because the first element, duty or a professional relationship, is lacking.

Duty: Establishing Standard of Care

The establishment of a professional relationship in most health care settings is created by the facility's contract to provide care to the client. In the workplace, we are concerned not with whether that relationship exists, but with what the provider should have done. The duty is to meet the standard of care or, put another way, to act like a reasonably prudent health care provider under the same or similar circumstances. The standard of care has been determined traditionally by evaluating a variety of sources including policies and procedures, professional practice acts, other state statutes and regulations, clinical guidelines promulgated by professional organizations, job descriptions, the literature, and expert witness opinions.

Could a critical pathway similarly be used as evidence of what should have been done for a patient, of what the patient should have done as a participant in his or her own care, of expected outcomes or expected time frames for patient outcomes? At the time of publishing this book, five states, including Maine, Vermont, Minnesota, Florida, and Maryland, have passed a law mandating a 5-year medical liability demonstration project. In these states, practice parameters (usually developed by a medical specialty society and may be endorsed by particular state agencies), including some as specific as critical paths, have force of law and may provide protection in the courts against malpractice suits. In Maine, this mandate means that physicians who have elected to be part of the project and follow guidelines established by an appointed committee are able, in a pretrial screening of claims, to assert an affirmative defense if they followed guidelines. As a result, the only issue becomes whether the professional met the standard, not whether the standard was appropriate. This has not yet been tested by the Maine courts. It is therefore extremely important that risk managers and others involved in development and implementation of critical pathways monitor state initiatives and the legal status of critical paths within the jurisdiction.

When a case is in litigation, once the plaintiff's attorney has reviewed the medical record, he or she will request copies of applicable policies and procedures, and then compare what the policy says should have been done with what is documented. Similarly, the attorney will compare the critical path with the treatment provided and the patient outcome. Discrepancies provide an opportunity to argue that noncompliance is evidence of malpractice. A lay jury will find this compelling.

Breach of Duty

In addition to establishing the duty, plaintiff must show that a breach of duty or violation of the standard of care occurred. Failing to use sterile technique when indicated, delegating specialized functions to untrained assistants, and ordering or administering an incorrect dosage of medication can all constitute a breach of duty.

Injury

A successful malpractice claim also requires that the plaintiff suffer an injury. Generally, this must be a physical injury such as paralysis, brain damage, fracture, or scarring. Emotional damages may be claimed in connection with the physical injury, but in malpractice actions in most jurisdictions, damages based on purely emotional injuries are prohibited.

Causation

The last element of malpractice, causation, can be the most problematic. There must be evidence that the provider's breach of duty caused the patient's injury. See Figure 9.1 for a case example.

The final analysis of the case example in Figure 9.1 is discussed in Figure 9.2.

The plaintiff was an 88-year-old female with congestive heart failure who was discharged to a home care agency following a brief hospitalization. The agency's documentation system included critical pathways with a charting by exception format. The critical pathway specified that the patient is taught to take her own pulse during visits 8-10. Although the nurse documented that the patient was not able to take her own pulse, the reason was never stated, nor was the variance explained in any detail. At the time of her deposition, the nurse was unable to recall the circumstances independently. Plaintiff's experts were prepared to testify that the nurse was negligent in not following the critical pathway and by not teaching the patient to take her own pulse. They argued that had the patient been properly educated, she would have noted bradycardia and arrythmias and would have notified her physician. Thus she would have survived the ensuing cardiac arrest.

FIGURE 9.1   Case example.

LIABILITY FOR DEVELOPERS

Example: A patient has brought an action against the hospital, physical therapist, and nurses who treated him, claiming that he received substandard care for the treatment of his leg fracture. In addition, plaintiff has named XYZ Committee (developers of the critical pathway), alleging that the critical pathway was inadequate, outdated, and did not include adequate physical therapy input, thus causing him to suffer a disability to his leg.

Whether there is a liability for the developers of the critical pathway will depend on the clinical accuracy, thoroughness of research, and documentation of the development process.

Who Will Develop the Critical Pathway?

Written policies and procedures should be established to guide this process and the individuals and committees involved.

- From a legal perspective, it is important that a committee be comprised of a representative from each discipline who may be involved in the client care process.
- Appropriate committees and departments, including risk management and legal counsel, should review the completed guidelines.
- Documentation of the education and experience (CV) of each reviewer should be retained. Utilize outside experts as necessary to ensure an appropriate level of review.
- The committee should document research tools utilized, including professional standards, practice parameters and guidelines, literature references, and other clinical/scientific sources.

Duty: A professional relationship existed by virtue of the agency agreement to provide services to the patient.

Breach of Duty: Failing to educate the patient in view of the critical pathway that indicates it should be done is not per se a breach of duty. A determinative factor in this case is the lack of documentation regarding why this was not done or followed up. For example, had the nurse documented that the patient stated she did not want to learn to take her pulse and refused to participate, and that repeat attempts to teach her were made, the case becomes far more defensible. Here, there is no evidence to support the deviation from the pathway, so the plaintiff's position gains credibility.

Injury: Cardiac arrest and death are physical injuries.

Causation: The issue is whether the failure to educate the patient caused her heart attack and death. Defendant's position in response to plaintiff's argument would be that even if the plaintiff had been taught to take her own pulse, she might not have called her physician in time, and even if she did, she would have suffered the cardiac arrest and died. Medical expert testimony from both sides would be heard by the jury.

FIGURE 9.2 Analysis.

- Evaluate whether the policies and procedures are compliant with JCAHO standards for documentation.

How Will the Critical Pathway Be Developed?

Written policies and procedures should guide this process as well. In the event that the validity or accuracy of the critical path itself is questioned, documentation of the development process will be crucial.

- Pilot studies should be considered to evaluate effectiveness. (Of course, the patient's physician should be consulted prior to including patient in pilot study.)
- The policy should address inservice of staff, specifically including the purpose of the critical pathway, resource people available, and orientation to policies and procedures.
- Retention of in-service records for staff.

In addition to clinical considerations, the following are important legally:

- Familiarize the staff with layout of forms that are part of the medical record.
- Definitions of all terms used.
- Identification of clinically appropriate patients.

- Identification of resource personnel.
- Clear explanation of documentation on the critical path with particular attention to documentation of variances.
- Purpose of the critical pathway: to improve patient care.
- Philosophy of critical pathways as means to cost-effectively provide and document care.
- Time frame for critical pathway, for example, preadmission, entire hospital stay, postdischarge.

How Will the Critical Pathway Be Evaluated?

Written policies and procedures indicating the evaluation process as it will be applied to critical pathways should be in place prior to the implementation of any critical pathway. The following considerations should be addressed:

- How variances will be monitored, analyzed, and addressed.
- The policy should address how outcome analysis will be analyzed and used to update the critical pathway.
- Who will conduct measurement outcomes.
- The frequency of critical pathway review, who will perform the review, and where and how this will be documented.
- Who will approve changes to the document as a result of variance analysis.
- How patient's input or evaluation will be used in the review process.
- Will information be released to outside agencies.
- To whom will data be distributed.

Once the policies and procedures have been developed, the team should document how each component of the policy was met for *each* critical pathway. A review of research literature should be conducted. The formulating committee should document all research tools, literature, and professional guidelines reviewed and implemented to formulate the critical pathway. It is important that the aspects of care that comprise the critical pathway are consistent with current research and clinical practice. Additionally, all the members of the multidisciplinary team should agree and approve the individual elements of the critical pathway. A pilot test can be performed to determine its effectiveness. The process used for the development of the critical pathway should be documented.

A worksheet to assist the developers in documentation should include the following information for each critical pathway:

- title of critical pathway;
- team members, including their names and departments;
- date and length of meeting;
- research performed to formulate the critical pathway;

- what literature was reviewed;
- what professional guidelines were reviewed;
- identify variances;
- identify how and where variances will be documented;
- identify who will perform outcome measurement;
- identify to whom the results of outcome measurement will be reported;
- identify how often the critical pathway will be reevaluated;
- identify any action taken to modify the critical pathway and the reason therefor;
- identify the in-service education plan and timetable.

The checklist in Figure 9.3 includes some crucial questions to consider during pathway development.

Even though all these steps have been followed, consideration should be given to providing a letter of indemnification or a statement regarding insurance coverage for individuals who participate in the development of the critical pathway (shifting the loss from the individual to the employer).

LIABILITY FOR USERS

Four situations in which the health care provider using a critical pathway could face potential liability have been identified:

- The provider is following the critical pathway but is negligent in the delivery of care;
- The provider is following the critical pathway, but the critical pathway dictates substandard care;
- The provider does not follow the critical pathway and fails to document the variance; and
- The provider misapplies the critical pathway.

The case example in Figure 9.4 is an illustration of a caregiver who follows the critical pathway, but is negligent in the delivery of care.

Figure 9.4 is a simple example of negligence in the prescribing and administering of medication that exists independently of the critical pathway. Any provider-patient relationship has the potential for liability when the four elements of malpractice (duty, breach of duty, injury, and causation) are met. The critical pathway as a tool for planning and documenting patient care will not protect the provider from his or her own act of negligence.

In the case outlined in Figure 9.5, a provider follows the critical pathway, but the critical pathway dictates a treatment that is not based on standard practice or research. Liability for the developer of this critical pathway would be likely,

Is it valid?	Did they lead to the health gains and projected costs?
Is it reproducible?	Are these recommendations similar to those produced by other teams for the topic identified?
Are they reliable?	Would another health professional interpret and apply the critical pathway in essentially the same way?
Are they a multidisciplinary process?	Do the guidelines formulated represent input from all disciplines involved in primary care for these patients?
Are the pathways clinically applicable?	Do the pathways conform to current research and clinical judgment and state to which population the pathways apply?
Are the pathways clinically flexible?	Do they identify the specifically known or generally expected exceptions to the recommendations?
Are the guidelines clear?	Do the pathways use unambiguous language and format that can be followed?
Are they updated?	Do critical pathways state when and how they are to be reviewed and when additional research will be conducted?
Have all these steps been documented?	Does the documentation include the participants involved and criteria used to measure outcomes?

FIGURE 9.3   Guidelines to avoid liability—a checklist for developers.

assuming that the pathway deviates from the standard of care. The user, however, has a right to rely, in good faith, on the plan developed, unless he or she knows or has reason to know that the standard of care is not being met. The critical pathway is not a substitute for a provider's professional judgment, and it will not protect the provider who blindly follows it. Nor will the institution escape liability when its critical pathways fail to meet the standard of care.

> *The provider is following the critical pathway,*
> *but is negligent in the delivery of care.*
>
> A nursing home resident was admitted to the hospital with a diagnosis of simple pneumonia. The pneumonia was institutionally acquired, and according to the critical pathway, cefotaxime 1 gm IV q8h and erythromycin 1 gm IV q8h were ordered by the treating physician. A nurse administered the medication, which had been prepared in the hospital pharmacy. The patient suffered an anaphylactic shock and, in fact, had a documented erythromycin allergy.

FIGURE 9.4 Case example.

> *The provider is following the critical pathway,*
> *but the critical pathway dictates substandard care.*
>
> A 40-year-old female admitted to a home care agency with a Stage II venous ulcer on her lower right leg. The critical pathway indicated the initial treatment to include applying aluminum hydroxide to the wound, allowing it to air dry, applying a gauze dressing that should be changed b.i.d. By the 3rd day of treatment, the ulcer had progressed to Stage III.

FIGURE 9.5 Case example.

Guidelines to Avoid Liability

- Ensure that critical pathways are developed by multidisciplinary teams that address all clinically relevant aspects of patient care.
- Provide for regular, periodic, documented review of all critical pathways against appropriate professional and clinical standards.
- Peer review of critical pathways should include outside consultants with specific clinical expertise.
- Carefully review critical pathways provided by or required by insurance companies or managed care organizations. These paths may have been developed with cost-effectiveness as a primary goal. Before you implement them, be certain that cost-effectiveness does not adversely impact quality of care.
- Be certain that critical pathways do not conflict with existing policies and procedures.
- As a provider, be familiar with the currently accepted treatments for your patients' conditions.
- If the critical pathway is clinically inappropriate or could cause harm to the patient, confer with the patient's multidisciplinary team members and agency administrators.

- Out-of-date or substandard critical pathways should immediately be suspended from use and reviewed.
- Have a mechanism in place to notify all health care providers that a particular critical pathway has been suspended and should not be used.

The case example in Figure 9.6 describes an incident where health care providers do not follow the critical pathway and then do not document resulting variances.

A failure to follow the critical path without documenting the variance causes communication breakdown between health care providers, potentially compromises patient care, and creates potential liability as well. A further problem arises when there is no follow-up action to a documented variance. One of the frequently cited dangers of the critical path is the "cookbook medicine" phenomenon. Once the critical path is implemented, it may be all too easy to follow and document without thinking.

Guidelines to Avoid Liability

- Each critical path should be accompanied by a policy and procedure detailing the proper way to document variances. Variances should be defined and the policy should specify where they should be documented.
- Sufficient space should be provided to document variances according to policy.
- All providers should be periodically educated to document variances properly.
- There should exist a consistent methodology for identifying and analyzing variances to determine individual occurrences vs. ongoing system problems.
- Ensure that appropriate action is taken and documented in response to identified variances. The policy should specify where to document follow-up on variances.

The provider does not follow the critical pathway and fails to document and follow up on the variance.

On a 4-day critical path for congestive heart failure, variances are to be noted as "P" for patient related, "S" for system related, and "C" for caregiver related. On day 1, the patient teaching was not completed because the nurse was not available when the patient was awake. On day 2, the patient teaching was not completed because the patient did not comprehend the instructions and the family member was unavailable. On days 3 and 4 again, the patient teaching was not done because the family member was not available. These variances were not appropriately documented, and it was not discovered that the teaching had not been done until the day of discharge.

FIGURE 9.6   Case example.

The case example in Figure 9.7 outlines the dangers associated with not tailoring the critical pathway to meet the needs of the patient.

Guidelines to Avoid Liability

- Develop a strategy for addressing comorbidity and common variances before critical pathways are implemented.
- Regularly review variance analysis data to determine the need for copaths or algorithms.
- Follow the written policies and procedures for development of critical pathways, including peer review and documentation, when developing copaths or algorithms.
- Each copath or algorithm should be accompanied by a written policy and procedure detailing how and when it should be implemented.
- Educate staff on availability of newly developed copaths and algorithms.

Critical Pathways and Documentation

The medical record has become an increasingly public document, subject to review by state surveyors, insurance carriers, managed care organizations, attorneys, and jurors as the situation dictates. Regardless of the system of documentation used—PIE, SOAP, FOCUS, charting by exception, or critical pathways, the information recorded in the medical record is critical evidence in a variety of legal proceedings, many of which involve health care providers as witnesses and sometimes as defendants.

Reference to the medical record as "your best friend or your worst enemy" in a malpractice action is nothing new to most health care providers. The medical record can prove to be a friend when it contains complete, legible documentation that indicates that standard of care was met. Records that reflect the care given in this way can greatly influence an attorney's decision whether to take on a case.

The provider misapplies the critical pathway.

A manic depressive patient was admitted to a psychiatric unit of a general hospital with a history of increasingly manic behavior over a 2-week period. The appropriate critical pathway was implemented. On days 3 and 4, the patient expressed suicidal ideation to staff; however, there was no mechanism to add a suicidality component to the existing critical pathway. The patient was discharged on day 5 without any documented suicidality assessment. Two days after discharge, she was readmitted via the emergency department following a suicide attempt.

FIGURE 9.7 Case example.

Likewise, well-documented records protect health care providers in the event suit is brought. Years may pass between the time a patient is cared for and the time the defendants become aware of potential liability. The medical record, however, is written at the time the events take place, while they are clear in the writer's mind. Medical record documentation is therefore highly credible and given great weight, despite a provider's inability to remember details of a case independently.

From a defense viewpoint, analysis of documentation of patient care by means of a critical pathway isn't much different from other methods of documentation. The issues typically include at least the following:

- Does the documentation accurately and completely reflect the care given?
- Is there evidence that the standard of care was met?
- Are patient assessments documented objectively?
- Are entries dated and timed when they are written?
- Are entries signed appropriately?
- Are entries legible?
- Is all patient teaching thoroughly documented?
- Has the record been altered in any way?

When used as a documentation tool, rather than merely as a guide to the patient's care, critical pathways can be designed to address these issues effectively. Critical pathways in conjunction with a charting by exception format can provide complete, accurate, and truthful documentation—the cornerstone of the defense of any malpractice case. Next, some important guidelines for designing critical pathways and documenting care on them will be discussed.

Documentation Issues

- Assessments should be documented objectively, indicating facts, observations, client's statements, and other measurable criteria. This is important particularly when narrative comments are required in order to note variances.
- Medical record entries are often used to determine the chronology of events; therefore, accurate dating and timing of variances is critical.
- All entries in the medical record should be properly signed. The critical pathway should incorporate a mechanism for identifying the writer of each entry. Initialing entries or a series of entries is legally acceptable provided there is a space for each writer to indicate his or her initials and corresponding signature.
- Space should be provided for staff to initial items (rather than check them off) for better accountability.
- Be certain that the patient's name appears on each page of his or her medical record.

- Accurately document all patient and/or family teaching. In malpractice litigation, health care providers often face claims that they provided insufficient information to patients regarding use of equipment, techniques for procedures, recognizing signs and symptoms, side effects of medications, and so on. One benefit of the critical pathway is that it incorporates documentation of an assessment of the patient's readiness to learn and the fact that teaching was done. Without supporting documentation, however, it's the client's word against yours. The content outline of the instructions given should be detailed in an accompanying policy to ensure that there is evidence of the information provided. It is also important to document specifically who in addition to the patient was educated.
- The patient should date and sign the pathway, indicating that he or she understood and received the instructions.
- Use proper abbreviations. Each agency should have an approved abbreviation list and a third party should be able to understand any patient record with reference to that list. Independently devising abbreviations can result in two problems. First, the abbreviation may be offensive to third parties. For example, a psychiatric clinic record indicated that a patient "looks horrible—TWO," the provider's own shorthand for "totally whacked out." Imagine the effect this information would have on a jury! A second problem is that nonapproved abbreviations may have duplicate meanings. For example, does "BS" mean bowel sound, breath sounds, or blood sugar? In litigation, it will be imperative to know what the writer intended.
- Properly correct mistakes in all parts of the record, including the critical pathway. A failure to do so could constitute an alteration of the record, diminishing the writer's credibility and rendering a case indefensible. Every agency should have a written policy on how to correct a mistake, add information, and make late entries, specifically with regard to information, contained on the critical pathway. While the law does not require documentation to be perfect, it does require corrections to be made according to policy.

Figure 9.8 shows the difference between subjective documentation (based on personal thoughts or feelings) and objective documentation (based in reality; having a scientific basis).

Critical Pathway Design Issues

- Any subjective terms used on a critical pathway must be objectively defined, either in the accompanying policy or on the path itself. For example, a variance code "P" that defined "patient related" could mean any number of things, including "patient refused treatment" and "patient not home for treatment." Likewise, if the assessment indicates "patient environment is safe," consider what you intend the word "safe" to mean and define it in a policy.

Subjective Documentation	Objective Documentation
(V) Wound appears infected	(V) Wound mottled with green, foul-smelling discharge
(V) Appears depressed	(V) Patient states, "I have never been so hopeless, I can't go on."
(V) Patient is uncooperative	(V) Patient refusing to do ROM exercises, and states, "I'm too tired."

FIGURE 9.8   Type of documentation.

- One of the benefits of the critical pathway is the active participation of the patient in his or her own care. If the patient is provided with a special version of the path, consideration should be given to type size (to avoid allegations that the print was too small to read); to terminology (to avoid allegations that it was too difficult to understand); and to eliminating any suggestion of guaranteed outcome.
- The use of a disclaimer for both the clinician's clinical path and the patient's should be considered. A general disclaimer should convey that the path provides a guideline for categories of cases, but that the staff recognize that each patient requires individualized care and that treatment decisions are dictated by the practitioner according to the patient's condition. A disclaimer on the patient's version of the path should indicate that the guidelines are provided to better inform the patient about his or her condition and treatment and suggest that every patient is different and therefore treatment and outcomes will differ.
- When designing the critical path, carefully consider the documentation that is expected to appear on the path itself. Legibility of documentation is an important concern, so determine whether enough space exists on the form for each piece of documentation required.
- Documentation of variances is the most crucial piece of charting on the critical path. It enables a third party to determine why something was not done and will certainly contribute greatly to the defense of cases in which the rationale for omitting certain care was frequently left to the imagination. Variances should be clearly defined on the critical path form itself, and sufficient space should be allotted for the associated explanatory documentation. Beware, however, that documentation specific to the cause of an event has traditionally been omitted from the medical record. For example, the fact that a test was not done due to short-staffing or broken equipment would traditionally be lim-

ited to an incident report. A critical pathway, however, might require this documentation as an explanation of a system or caregiver variance. The documentation could be done on the pathway or on a separate variance report. The implications of discovery rules in each jurisdiction could influence the most prudent approach to the documentation of variances; therefore, facility risk managers and attorneys should be consulted.

- Before critical paths are implemented, written policies should be developed to address the philosophy of the critical path and how to complete each portion of the path. Inservice should be provided to all staff prior to implementation of the critical path as a documentation tool.
- Critical paths often contain much of the documentation concerning the patient's course of treatment and, as such, are an important part of the medical record. If paper records are the primary format in your institution, the medical records department should be consulted during the design phase to determine how the path will best be stored in the medical record. A certain paper size, paper type, or folding pattern may best suit the needs of the facility and enable better storage and duplication of this important information.

  If computerized patient record systems are utilized, only authorized users will have access to patient records and only to those portions of the patient records relevant to their function. Patient confidentiality must be maintained.
- A computer-based patient record system should include a security system that, as far as is practical, permits only authorized users to access patient records and permits authorized users to access only those portions of the records that are relevant to their particular functions. The security measure used can be as simple as a password or as complex as voice activation or thumbprint activation.

CONCLUSIONS

The examples used in this chapter were hypothetical because the law has not yet been tested with regard to critical pathways. The critical pathway may carry potential liability for developers and users alike, and the examples in this chapter are meant to heighten the reader's awareness. However, liability may be avoided by establishing and adhering to written policies and procedures that document the development, implementation, and evaluation of the critical pathway. Following the critical pathway is not a substitute for using your professional judgment. Developers and users of the critical pathway should be aware of any legislation pending in their states that affects the legal status of the critical path and how it may be used for or against the professional in a liability claim.

BIBLIOGRAPHY

Ballard, D., & Cohen, J. (1995). Confidentiality of patient records in the computer age. *Journal of Nursing Law, 4*(2), 49.

Dick, R. S., & Stein, E. B. (Eds.). (1991). *The computer based patient record: An essential technology for health care.* Washington, DC: National Academy Press, Institute of Medicine.

Duff, L. A., Kitson, A. L., Geers, K., & Humphris, D. (1996). Clinical guidelines: An introduction to their development and implementation. *Journal of Nursing Administration, 23,* 887–895.

Black, H. C. et al. (Eds.). (1983). *Black's law dictionary* (5th ed.). St. Paul: West.

Goodwin, D. (1992). Critical pathways in home health care. *Journal of Nursing Administration, 22*(2), 35–40.

Jamanow, J. K. (Ed.). (1995). Do guidelines reduce malpractice lawsuits? *Healthcare Standards Update, 6*(1), 4.

Kapp, M. (1994). The clinical practice parameters movement: The risk manager's role. *Journal of Health Care Risk Management, 2,* 33–38.

Mitchell, S., Price, C., Schneider, M., Ferrara, K., & Youngberg, B. (1995, December 1). Legal issues and practice guidelines. Managing the risks of managed care. *Home Health Digest* (Newsletter), (9), 1–8.

Powell, B. (Ed.). (1994). Critical paths can diminish risks, increase accountability. *Hospital Risk Management, 16*(2), 17–28.

Roman, K. (1995). Practice policies: Potential implications for malpractice litigation. *Journal of Healthcare Risk Management, 5*(3), 37–45.

Schanz, S. J. (Ed.). (1996). Parameters, guidelines and protocols. *Legamed, 2,*(4), 1–8.

Solomon, R. P. (Ed.). (1995). Practice guidelines and critical pathways; Swords or shields in malpractice litigation? *Hospital Risk Control, 6,* 1–6.

Waller, A. (1991). Legal aspects of computer based patient records and record systems. In R. S. Dick & E. B. Steen (Eds.), *The computer based patient record: An essential technology for health care* (App. B, p. 156). Washington, DC: National Academy Press, Institute of Medicine.

10

Critical Pathways and Computerization: Issues Driving the Need for Automation

*Debra A. Slye**

DEMANDS OF MANAGED CARE: BALANCING QUALITY AND COSTS

In an attempt to meet the demands of managed care, systems of health care delivery are in the process of evolving to support population-based care across the continuum (see Figure 10.1). Health care enterprises today are faced with various internal and external incentives aimed at the delivery of improved quality patient care at a reduced cost. In the absence of well-defined indices of quality, cost-cutting becomes the management target. At the present time, most enterprises are not equipped with the data necessary to determine which clinical processes and interventions are the most cost-effective in achieving the desired outcomes. To better manage and drive the care process, we need quality information. The volume of data required to develop quality information can be exhaustive, and the transformation of that data to information can be overwhelming if relegated to a manual process. Integrating an automated approach is essential to make the process realistic and the resulting information applicable.

*The author wishes to acknowledge Barbara F. Hunstein, M.S.N., R.N. (in memoriam), for her pioneering work in the automation of clinical pathways and her contributions to this chapter.

PAST	$\rightarrow$	PRESENT
Acute inpatient care		Continuum of care
Treating illness		Maintaining wellness
Individual patient		Accounting for health status of populations
Fragmentation		Consolidation
Fee-for-service		Fixed, prepaid rates for population
Revenue-based		Cost/quality, population-driven
Patients		Covered lives
Fill beds–excess capacity		Treat at appropriate care level
Closed systems		Open systems

FIGURE 10.1   Evolution in Care Delivery.

Note. Adapted from *Health care's new information age: Moving from concept to reality,* by D.I. Becker and C. Hudson, 1994, Baltimore: Alex, Brown.

MANAGEMENT OF PATIENT CARE ACROSS THE CONTINUUM

Case management is a strategic approach to the management of cost and quality outcomes across the continuum. The overall goal of case management is to move patients proactively through the care process by monitoring expected outcomes, removing barriers to progress, and ensuring appropriate utilization of resources. Most case management approaches involve the use of standardized critical pathways as a method of enhancing care efficiency and promoting adherence to standards for a given patient population. Experiences in non-health-care-related industries has shown that definition of a "best" practice model and adherence to that model reduce unintended variations in practice for a given population. Reductions in variation have been associated with improved outcomes and reduced costs.

Critical pathways are multidisciplinary plans of care that outline the typical course for approximately 80% of the cases of a particular type or diagnostic category incorporating critical events and key interventions that must occur in a given time frame and sequence to achieve specific patient outcomes within a predetermined length of stay. Critical pathways have become an accepted method of guiding clinical practice for individual patients. Through the cross-patient analysis of data extracted from the pathway (care delivered, outcomes achieved, variances experienced), pathways can serve as a tool for coordinating the care of patient populations (Crummer & Carter, 1993). In many care settings, critical pathways have been shown to improve interdisciplinary interaction, encourage timely interventions, improve documentation of care delivered, reduce redundancy in documentation, enhance analysis of outcomes, facilitate concurrent quality improvement, reduce length of stay, and contain costs (Hunstein, 1995). For the critical pathway to be useful in handling clinical data, managing the care process, documenting resource utilization, and analyzing cost-effectiveness, information systems are required (Crummer & Carter). If continuum-focused, pathways can be used to guide health maintenance and manage disease states in an effort to reduce costly hospitalizations (Lumsdom & Hagland, 1993).

ADVANTAGES OF AUTOMATING PATHWAYS

The majority of critical pathways in use today are paper based. Once a standard pathway is designed on paper, it can be applied to an individual patient. However, the degree to which a paper-based pathway can be customized to an individual patient is severely limited. Paper pathways are often used as a reference document, not an interactive tool to truly manage the care of individual patients. While some enterprises have found creative layouts that support the documentation of care and variances using a paper-based pathway, the data from such paper documentation must still be manually extracted to accomplish the variance analysis necessary to make critical decisions. Some enterprises have gone to the extent of developing scan forms for the acquisition of variances. However, the care documentation process is usually a separate process.

Automating the pathway process greatly enhances the ability to customize a standard pathway to meet the needs of individual patients. Advantages to automating pathways are as follows:

- Timely and accessible information
- Shared data elements eliminates redundant data entry
- Improved information acquisition with integration of various systems
- Order entry can occur directly from the pathway
- Results/status can be reported to the pathway
- Documentation can be performed against orders within the pathway
- Prompts, scheduled activity lists, and conditional flags are generated by the system
- Multidisciplinary involvement in the planning process is improved
- Integrated views of information improve decision making

Incorporating the real-time patient care documentation as an integral component of the automated pathway brings the standard pathway to the core of the care process, facilitates the care provider's awareness of and adherence to the standard pathway, and automatically generates variances based on the presence or lack of care documentation. Ideally, the automated pathway should be integrated with the orders management application so that all disciplines are accommodated. As with care documentation, integration of orders into the pathway focuses each care provider on the standard and allows him or her to consider how a particular patient is adhering or deviating from that standard. If a patient is experiencing variances, rationale for these variances can be documented and analyzed to determine whether such variances might be avoidable, conditions causing such variances might be resolvable, or whether the variance is truly patient-specific and does not warrant any further consideration from a standards perspective.

Before automating the process, it is essential to define the demands for information within multiple care settings and those that support the linkages between care settings. Access to an automated system is essential for maintaining continuity of patient care information across the continuum (Patterson, Blehm, Foster, Fuglee, & Moore, 1995).

CRITICAL PATHWAY DEVELOPMENT

The first stage of pathway development involves the analysis of current practice. Analysis of aggregated patient data is essential to minimize sampling error and ensure statistical validity. While individual patient data is the input, the focus needs to be on the global results of processes rather than focusing on individual patient outcomes. To observe any significant improvements in quality, efficiency, and cost-effectiveness, process reengineering must be directed at the population level. In the absence of an automated process, data acquisition can be tedious and analysis time consuming.

Automating at the Point of Care

The first step of the automated process is to acquire data from the point of care ideally as a by-product of care documentation. Capturing data from their original source wherever possible is a critical component, whether this be the physician, nurse, other health care provider, patient/family, or from a physiologic monitor or other bedside device. Data acquired from the point of care can be stored for later abstraction, summarization, and aggregation on a cross-patient basis to describe current practice and common variations in practice for a specific diagnostic category or other defining premise and sorted by care provider. Data from current practice can be analyzed to determine which route achieves the best outcomes clinically and financially.

Charting by exception is a common methodology employed to reduce the need for care documentation. Charting by exception has been extended in some cases to exception reporting, whereby only those items that occur in other than the standard manner are documented. Using such methods, the lack of documentation assumes that the patient's presentation was within normal limits, orders and interventions were carried out according to the standard, and outcomes were met as predicted. This extensive approach to charting by exception has not always been favorably received by the risk management and legal community. Without adequate documentation of care that was actually carried out and the associated patient responses, it is sometimes difficult to recreate retrospectively an actual patient scenario.

Automating patient care documentation expedites data entry and enables a legally more acceptable approach to charting by exception. Through the automated process, care providers can be presented with the enterprise-defined normal values for an assessment finding or intervention response, or be permitted to pull over the values that were last entered in the event that nothing has changed. This approach to charting by exception provides a more complete record and a better audit trail as to what actually occurred without bogging down the data entry process.

Infusion of Scientific Evidence

Once best practice is defined based on current practice, review of current research is necessary to validate the clinical appropriateness of the model. Without the incorporation of scientific evidence, pathways can become purely anecdotal. Current practice could also be correlated with accepted national standards or other benchmark data. Ultimately, a "best" practice pathway is built to guide the management of individual patients. As with any quality improvement effort, implementation of pathways is a continuous process involving the ongoing analysis of variances and refinement of pathways to meet cost and outcome goals.

Resource Utilization

The critical pathway can be used as a cost and resource allocation tool. The resolution of problems as evidenced by the achievement of goals and expected outcomes supply information regarding its clinical effectiveness, and the variance analysis becomes the foundation for quality improvement (Lumsdom & Hagland, 1993). A variety of tools to estimate resource utilization are currently available in the market. However, their completion is commonly a secondary process to care documentation and, therefore, often not clearly reflective of the actual care delivered. Ideally, resource consumption is automatically acquired as a direct result of care documentation so that it most accurately reflects the actual care delivered. Tying a relative value to an intervention multiplied by the level of expertise of the provider delivering the care is one mechanism of defining resource utilization. Calculating resource consumption by pathway supports forecasting of resource requirements. Analysis of variances in resource consumption and associated outcomes by pathway may facilitate refinement of the pathway to optimize the cost-effectiveness of care.

Costs of Care

To ensure success in contract negotiation, health care enterprises must implement better mechanisms of estimating the actual costs of delivering care for specific diagnoses, procedures, or pathways. Traditional systems have used patient charges to estimate costs of care. It is well known that charges do not equal costs. With the onset of capitation (i.e., negotiation of a fixed price per covered life), charges become a mute point and the value of obtaining true costs of care escalates. Similar to resource utilization, costs of care can be extracted by the system based on care documented, the salary level of the care provider, and the cost of any associated supplies. By costing out the pathway, some enterprises have been able to offer a fixed price product for certain diagnostic categories (Hunstein, 1995).

IMPLEMENTATION OF CRITICAL PATHWAYS

As consistently stated in earlier chapters, multidisciplinary involvement in the development of pathways is essential to the successful incorporation of pathways into the overall care process. With the availability of interactive pathways, the pathway becomes a working tool to guide the care of a particular patient. The overall flow of an automated pathway is illustrated in Figure 10.2. Upon entry of an assessment finding or diagnosis, the care provider can be presented with a list of pathway options. In addition, a patient may comply with specific criteria or fall into certain risk groups for which an automated system can provide alerts or conditional branching of the pathway to accommodate such characteristics. Incorporating rules-based logic into pathways facilitates customization to meet the needs of the patient.

A critical pathway is typically laid out with an element of time across the horizontal axis and aspects of care along the vertical axis (see Figure 10.3). Aspects of care may include physician orders, nursing interventions, ancillary services, consultations, and expected patient outcomes. Automating pathways permits varying views based on care provider role and privilege level. Higher level views of the pathway may be provided to physicians and case managers, for example, while comprehensive views may be provided to direct care providers.

Discipline-specific views streamline the display to only the salient aspects of the pathway based on discipline requirements. For example, physical therapists may be primarily interested in viewing physical therapy orders, occupational therapy orders, nursing interventions and expected outcomes related to the patient's physical mobility, strength and endurance, as well as self-care status. In some cases, physical therapists may want to have access to the medication administration record to view when the last pain medication was delivered. An automated system must be flexible enough to accommodate the needs of all disciplines.

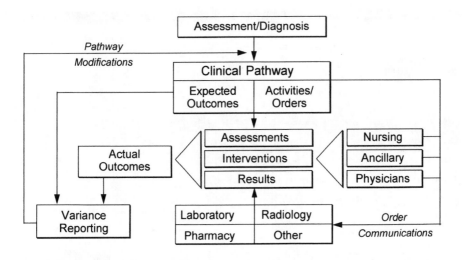

FIGURE 10.2   Automated pathway flow diagram.

Note: Copyright 1996. Space Labs Medical, Inc. Used with permission.

Orders and activities that comprise the critical pathway (e.g., assessments, interventions, diagnostic tests, and treatments) can be electronically communicated to the responsible parties. This communication could include the posting to associated work schedules or activity lists for the various care providers. Once a result is made available to the system, it can be communicated back to the care provider or incorporated into the appropriate flow sheet as specified by the health care enterprise.

When evaluating the patient's achievement of expected outcomes, the care provider should be presented with the salient data in the system needed to make that determination. Upon review of this clinical information, the care provider can make a judgment as to whether the outcome has been met, needs to be reevaluated at a later date, or whether the substance of the outcome itself requires adjustment.

Whenever an aspect of care has not been documented and is past due, the system can generate a flag alerting the care provider that it is overdue. When the overdue conflict is resolved, the care provider can be prompted to document whether there was a variance to the standard pathway and the associated rationale for the variance. Similarly, variances can be documented against outcomes that are not met according to the time frame specified in the standard pathway.

DECISION SUPPORT

A key purpose for the automation of clinical information is to facilitate the assimilation of data to enhance fast, effective decisions. Decision support can range from

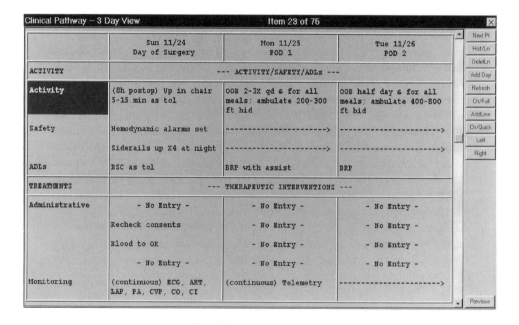

FIGURE 10.3 Screen capture of automated pathway.

Note: Used with permission from Space Labs Medical, Inc.

alerts, prompts, and reminders to complex, rules-based algorithms and clinical protocols. "Unaided humans are not capable of providing the persistent commitment to detail and to decision making logic (rules) necessary to effect standardization of care comparable to that achieved by an executable computerized protocol" (East et al., 1995, p. 244).

Alerts, Prompts, and Reminders

Alerts may be incorporated into an automated system to notify the care provider of a critical event that has relevance to the current function. Flagging of clinical values that are out of range and those that are critically abnormal expedites notification of potentially significant changes in a patient's condition. Prompts may be used to guide the care provider through the data entry process to ensure complete, consistent, and valid information. Reminders may include conditions and recommended actions such as the most recent laboratory test result that has bearing on a new order, suggested supplemental orders, such as assessment of kidney function when ordering nephrotoxic antibiotics, or contraindications to orders based on the patient's condition or existing therapies, and prompts for suggested alternatives (Overhage, Mamlin, Warvel, Tierney, & McDonald, 1995).

Algorithms

An automated algorithm can be considered a decision tree with a series of yes/no questions with branching logic that serve to guide diagnosis and patient management. As defined by the Agency for Health Care Policy and Research (McCormick, Moore, & Siegel, 1994, pp. 96–97), algorithms have the following advantages:

- Result in faster learning, higher retention, and better compliance with established practice standards.
- Provide structure for retrospective quality review activities.
- Identify situations in which testing is unnecessary. The only tests required are those on which management strategies are dependent.
- Permit the testing required to explore the impacts of changing assumptions about outcomes, costs, and preferences on the structure and content of clinical guidelines.

When automating algorithms, it is essential to eliminate all ambiguity from the decision steps and rules to ensure universal application.

Clinical Protocols

Clinical protocols are defined as a series of care requirements for a particular treatment or therapy. Common in the complex critical care environment and to support the triage process in emergency departments, protocols often include rules that support decisions, for example, ventilator weaning, anticoagulation management, or acute asthma triage. Similar to critical pathways, protocols are generally built based on aggregated data of patient responses and associated interventions that result in optimal outcomes.

Clinical protocols can be considered minipathways in themselves or incorporated as part of a larger pathway. Using an automated approach, rules can be integrated so that the care provider is only presented with the interventions that apply based on the specific patient's presentation. For example, a pneumonia pathway may have an inherent branch that accounts for temperature status (i.e., normothermic or hyperthermic). The protocol rules may define a set of interventions that apply when the patient's temperature is greater than 102°F. If the patient's temperature exceeds 102°F, the care provider could be presented with the hyperthermic branch of the pneumonia pathway. All relevant orders could be automatically sent to the associated departments, posted to the requisite work lists, and all relevant documentation screens made available. The degree to which this process occurs automatically can be based on an enterprise's preference. In this manner, compliance with the "best practice" standard is enhanced.

Merging Pathways and Protocols

In today's health care climate, many patients are presented in very acute states, often with multiple comorbidities—the care for which may constitute a pathway in itself. Without automation, customizing the pathway to the individual patient and merging relevant pathways can be an overwhelming task. Using an automated method not only facilitates the building of standard pathways, but allows for the ongoing customization of the pathway for the individual patient and the merging of multiple pathways and protocols as the patient's condition warrants.

Case Scenario

A patient with degenerative joint disease may be scheduled for a total hip arthroplasty. The enterprise has developed a standard pathway for total hip arthroplasty. However, this patient also has chronic diabetes mellitus. Due to the large volume of diabetic patients seen at the enterprise, a standard hybrid pathway for total hip arthroplasty with chronic diabetes mellitus was developed and selected for this patient. Postoperatively, he experiences a myocardial infarction necessitating the suspension of the majority of the total hip arthroplasty orders and interventions and initiation of the acute myocardial infarction pathway. The system presents the care provider with the orders, interventions, and expected outcomes related to the total hip arthroplasty for possible suspension or alteration. The care provider is prompted to provide rationale for the variances. The pathway aspects related to diabetes may be continued as originally planned or adjusted as necessary based on the patient's current condition. When adding the myocardial infarction pathway, the system can automatically screen for conflicts or overlaps between the existing orders and interventions. Based on an enterprise's definitions, the system can be configured to resolve obvious conflicts between pathways and present the care provider with questionable conflicts. Aspects of the total hip pathway can be resumed as appropriate.

OUTCOMES MANAGEMENT AND VARIANCE ANALYSIS

The principles for dealing with outcomes management and variance analysis are consistent whether automated or paper pathways are used. (See chapter 4 for review of outcomes management and variance analysis.) The greatest differences lie in the quantity of data that may be collected using automated pathways and the usefulness of that data. Due to the difficulty of data collection with paper pathways, there is little documentation of the effectiveness of this data from an outcomes perspective. On the other hand, to date, researchers have documented that automated critical pathways have been shown to improve the following (Hunstein, 1995; Morrall, 1996; Remmlinger, Ault, & Hanrahan, 1995):

- organization and timeliness of care through the multidisciplinary use of an automated clinical information system
- care provider compliance with clinical practice standards as evidenced by quality improvement reviews and variance reporting
- decrease length of stay of up to 25%
- control hospital costs, reduce inpatient Medicare losses by more than 50%
- increase time spent in direct patient care
- improve documentation of care against orders by 29% over manual charting
- improve documentation of outcomes and variances from expected

Impediments to Data Analysis

Although the impediments to data analysis from paper pathway data are many, measurement of clinical practice and outcomes with automated pathways has been precluded by lack of standards in terminology related to patient care delivery (assessment findings, interventions, outcomes). Nomenclature is inconsistent within and among disciplines, thus making it difficult to quantify current practices and benchmark optimal care. To support maximal data integration, terminology needs to be truly multidisciplinary. This may require substantial give and take from all disciplines, but the process is essential to the success of data retrieval and comparison. See Figure 10.4 for a listing of current standards efforts.

Standard and Sponsor	Type of Standard
Nursing Minimum Data Set (Werley & Lang, 1988)	Patient demographics, patient responses, interventions, patient outcomes, resource consumption/nursing intensity
North American Nursing Diagnosis Association (NANDA) (NANDA, 1994)	Nursing diagnoses by human response patterns
Nursing Intervention Classification (NIC): Iowa Interventions Project (McCloskey & Bulechek, 1992)	Nursing interventions, patient outcomes
Omaha Community Health System (Martin & Scheet, 1992)	Nursing diagnoses, problem classification scheme, interventions, patient outcomes
Home Health Care Classification (Saba, 1992)	Nursing diagnoses, interventions, discharge status
Nursing Intervention Lexicon and Taxonomy (Grobe, 1992)	Nursing interventions in community health
National Library of Medicine (NLM) (NLM, 1991)	Unified Medical Language System Unified Nursing Language System (proposed)
International Classification of Disease - Clinical Modification (ICD9-CM) (ICD9-CM, 1992)	Diseases, factors influencing health status, external causes of injury, diagnostic and therapeutic procedures
Diagnostic and Statistical Manual of Mental Disorders (DSM-III-R) (American Psychological Association, 1992)	Mental disorders
Systematized Nomenclature of Medicine (SNOMED III) (College of American Pathologists, 1992)	Patient findings (anatomic and morphologic references, living organisms, signs and symptoms, nursing diagnoses, medical diagnoses, administrative, therapeutic, diagnostic and nursing procedures, social conditions)
Physician's Current Procedural Terminology (CPT) (American Medical Association, 1991)	Procedures by service performed by physicians
International Classification of Clinical Services (ICCS) (Mendenhall, 1987)	Procedures (limited domain, greater specificity than SNOMED III)
Laboratory Observation Identifier Names and Codes (LOINC) (Regenstrief Institute and LOINC Committee, 1995)	Names and codes for laboratory test results
National Council for Prescription Drug Programs (NCPDP)- National Drug Code (NDC)	Medications
World Health Organization (WHO) Drug Codes	Medications (more extensive than NDC)

FIGURE 10.4   Current standards efforts.

Note. Standards for medical identifiers, codes, and messages needed to create an efficient computer-stored medical record. *Journal of the Medical Informatics Association, 1*(1), 1–7, by Board of Directors of the American Medical Informatics Association, 1994. The nursing minimum data set: Use in the quality process. *Journal of Nursing Care Quality, 10*(1), 9–15, by A. Coenen and D. Schoneman, 1995. Terms used by nurses to describe patient problems: Can SNOMED III represent nursing concepts in the patient record? *Journal of the American Medical Informatics Association, 1*(1), 54–61, by S. B. Henry, C. R. Holzemer, and K. E. Campbell, 1994.

CONCLUSION

In summary, the adoption of managed care has necessitated a change in the way we think about bringing care and services to the patient. Focusing on improving the quality and standardization of care has been shown to reduce costs and enhance patient satisfaction. Innovative solutions are available to bring technology to the point of care, facilitate cross-training, and promote efficient decision making. Data are essential to evaluating current processes for opportunities to improve both clinical outcomes and costs.

Automated, integrated clinical information solutions that incorporate multidisciplinary critical pathways have been shown to guide clinical practice efficiently and effectively toward the achievement of desired patient outcomes. Clinical information systems that expedite the delivery and documentation of patient care from a multidisciplinary perspective according to expected clinical outcomes positively contribute to quality improvement and support cost-containment efforts. Automated systems facilitate managing the care process and analyzing variances to enhance outcomes, patient satisfaction, and overall system efficiency.

REFERENCES

American Medical Association (1991). *Physician's current procedural terminology.* Chicago: American Medical Association.

American Psychological Association (1992). *Diagnostic and statistical manual of mental disorders.* Washington, DC: American Psychological Association.

Becker, D. I., & Hudson C. (1994). *Health care's new information age: Moving from concept to reality.* Baltimore: Alex, Brown.

Board of Directors of the American Medical Informatics Association (1994). Standards for medical identifiers, codes, and messages needed to create an efficient computer-stored medical record. *Journal of the American Medical Informatics Association, 1*(1), 1–7.

Coenen, A., & Schoneman, D. (1995). The nursing minimum data set: Use in the quality process. *Journal of Nursing Care Quality, 10*(1), 9–15.

Crummer, M. B., & Carter, V. (1993). Critical pathways-The pivotal tool. *Journal of Cardiovascular Nursing, 7*(4), 30–37.

East, T. D., Wallace, C. J., Franklin, M. A., Kinder, T., Sailors, R. M., Carlson, D., Bradshaw, R., & Morris, A. H. (1995). Medical informatics academia and industry: A symbiotic relationship that may assure survival of both through health care reform. *American Medical Informatics Association, 2,* 243–247.

Henry, S. B., Holzemer, C. R., & Campbell, K. E. (1994). Terms used by nurses to describe patient problems: Can SNOMED III represent nursing concepts in the patient record? *Journal of the American Medical Informatics Association, 1*(1), 54–61.

Hunstein B. (1995). From beta site to hospitalwide clinical pathways. *Healthcare Informatics, 12*(2), 34–35.

International Classification of Disease - Clinical Modification (9th ed.). (1992). Salt Lake City, UT: Med-Index.

Lumsdom, K., & Hagland, M. Mapping care. (1993). *Hospitals & Health Networks, 67,* 34–40.

Martin, K. S., & Scheet, N. (1992). *The Omaha system: Applications for community health nursing.* Philadelphia: Saunders.

McCloskey, J., & Bulechek, G. (Eds). (1992). *Nursing interventions classification (NIC).* St. Louis: Mosby Year Book.

McCormick, K. A., Moore, S.R., & Siegel, R.A. (1994). *Clinical practice guideline development: Methodology perspectives.* Rockville, MD: U.S. Department of Health and Human Services.

Mendenhall, S. (1987). The ICCS code: A new development for an old problem. In W. W. Stead (Ed.), *Proceedings of the Eleventh Symposium on Computer Applications in Medical Care.* (pp. 703–709). IEEE Computer Society Press.

Morrall, K. (1996). Paving the road to Medicare savings. *Hospitals & Health Networks, 70*(2), 43.

National Library of Medicine (1991). *Fact sheet: Unified medical language system.* Bethesda, MD: U.S. Department of Health and Human Services, Public Health Service, National Institutes of Health.

North American Nursing Diagnosis Association (NANDA). (1994). *Nursing diagnosis: Definition and classification, 1995–1996.* Philadelphia: NANDA.

Overhage, J. M., Mamlin, B., Warvel, J., Tierney, W., & McDonald, C. J. (1995). A tool for provider interaction during patient care: G-CARE. *American Medical Informatics Association,* 178–182.

Patterson, P. K., Blehm, R., Foster, J., Fuglee, K., & Moore, J. (1995). Nurse information needs for efficient care continuity across patient units. *Journal of Nursing Education, 25*(10), 28–36.

Regenstrief Institute and Laboratory Observation Identifier Names and Codes (LOINC). (1995). *Laboratory observation identifier names and codes: Users guide vs. 1.0.* Indianapolis, IN: Regenstrief Institute.

Remmlinger, E., Ault, S., & Hanrahan, L. (1995). Information technology implications of case management. *Healthcare Information Management, 9*(1), 21–28.

Saba, V. K. (1992). The classification of home health care nursing diagnoses and interventions. *Caring, 11*(3), 50–57.

Werley, H., & Lang, N. (Eds). (1988). *Identification of the nursing minimum data set.* New York: Springer.

BIBLIOGRAPHY

Abbott, J., Young, A., Haxton, R., & Van Dyke, P. (1994). Collaborative care: A professional model that influences job satisfaction. *Nursing Economic$, 12*(3), 167–169, 174.

Ahrens, T. (1992). Nurse clinician model of managed care. *AACN Clinical Issues in Critical Care, 3*(4), 761–768.

American Nurses Association. (1988). *Nursing Case Management* (Publication No. NS-32). Kansas City, MO: American Nurses Association.

Barnsteiner, J. H., Mohan, A., & Milberger, P. (1992). Implementing managed care in a pediatric setting. *AACN Clinical Issues in Critical Care, 3*(4), 777–784.

Becker, D. I., & Hudson, C. (1994). *Health Care's New Information Age: Moving from Concept to Reality.* Baltimore: Alex, Brown.

Board of Directors of the American Medical Informatics Association. (1994). Standards for medical identifiers, codes, and messages needed to create an efficient computer-stored medical record. *Journal of the American Medical Informatics Association, 1*(1), 1–7.

Brennan, P. F., & Fitzpatrick, J. J. (1992). On the essential integration of nursing and informatics. *AACN Clinical Issues in Critical Care Nursing, 3*(4), 797–803.

Capuano, T. A. (1995). Clinical pathways: Practical approaches, positive outcomes. *Nursing Management, 26*(1), 34–37.

Coalition for Critical Care Excellence (1995). *ICU cost reduction: Practical suggestions and future considerations: Vol. 2. Personnel.* Anaheim, CA: Society of Critical Care Medicine.

Coenen, A., & Schoneman, D. (1995). The nursing minimum data set: Use in the quality process. *Journal of Nursing Care Quality, 10*(1), 9–15.

Crummer, M. B., & Carter, V. (1993). Critical pathways - The pivotal tool. *Journal of Cardiovascular Nursing, 7*(4), 30–37.

DelTogno-Armanasco, V., Olivas, G. S., & Harter, S. (1989). Developing an integrated nursing care management model. *Nursing Management, 20,* 26–29.

East, T. D., Wallace, C. J., Franklin, M. A., Kinder, T., Sailors, R. M., Carlson, D., Bradshaw, R., & Morris, A. H. (1995). Medical informatics academia and industry: A symbiotic relationship that may assure survival of both through health care reform. *American Medical Informatics Association, 2,* 243–247.

Fields, M. (1994). Critical pathways: High roads to better patient care. *Healthcare Informatics, 11*(8), 40–44.

Gage, M. (1994). The patient-driven interdisciplinary care plan. *Journal of Nursing Administration, 24*(4), 26–35.

Graybeal, K. B., Gheenm, M., & McKenna, B. (1993). Clinical pathway development: The Overlake model. *Nursing Management, 24,* 42–45.

Gunderson, L., & Kenner, C. (1992). Case management in the neonatal intensive care unit. *AACN Clinical Issues in Critical Care, 3*(4), 769–776.

Hunstein, B. (1995). From beta site to hospitalwide clinical pathways. *Healthcare Informatics, 12*(2), 34–35.

Joos, L. J., Thomson, C., & Weber, A. (1995, Jan./Feb.). The development of case management in cardiac services: Creating change for future survival. *Journal of Cardiovascular Management,* 27–32.

Lumsdom, K. (1993). Clinical paths: A good defense in malpractice litigation? *Hospitals & Health Networks, 68*(6), 58.

Lumsdom, K., & Hagland, M. (1993). Mapping care. *Hospitals & Health Networks, 67,* 34–40.

Mamaril, M. (1995). Defining data elements: A joint venture in establishing data bases to support clinical perioperative nursing practice. *Breathline, 15*(4), 11, 14.

Martich, D. (1993). The role of the nurse educator in the development of critical pathways. *Journal of Nursing Staff Development, 9,* 227–229.

McCormick, K. A., Moore, S. R., & Siegel, R. A. (1994). *Clinical Practice Guideline Development: Methodology Perspectives.* Rockville, MD: U.S. Department of Health and Human Services.

McElroy, M. J., & Campbell, S. (1992). Case management with the nurse manager in the role of case manager in an interventional cardiology unit. *AACN Clinical Issues in Critical Care, 3,* 150.

McFarland, M. (1995). Knowledge engineering of expert systems for nursing. *Computers in Nursing, 13*(1), 32–37.

Meyer, J. W., & Feinfold, M. G. (1995, February). Integrating financial modeling and patient care reengineering. *Healthcare Financial Management* (Newsletter), 33–40.

Morrall, K. (1996). Paving the road to Medicare savings. *Hospitals & Health Networks, 70*(2), 43.

Myrick, J. A., & Irby, S. P. (1995). Rethinking health care reform. *Healthcare Information Management, 9*(1), 9–20.

Nelson, M. S. (1993). Critical pathways in the emergency department. *Journal of Emergency Nursing, 19*(2), 110–114.

Overhage, J. M., Mamlin, B., Warvel. J., Tierney, W., & McDonald, C. J. (1995). A tool for provider interaction during patient care: G-CARE. *American Medical Informatics Association,* 178–182.

Ozbolt,. J. G., & Graves, J. R. (1993). Clinical nursing informatics: Developing tools for knowledge workers. *Nursing Clinics of North America, 28*(2), 407–425.

Patrick, M. S. (1992). Benchmarking - Targeting "best practice." *Healthcare Forum, 35*(4), 71–74.

Patterson, P. K., Blehm, R., Foster, J., Fuglee, K, & Moore, J. (1995). Nurse information needs for efficient care continuity across patient units. *Journal of Nursing Administration, 25*(10), 28–36.

Rapoport, J., Teres, D., Lemeshow, S., & Gehlbach, S. (1994). A method for assessing the clinical performance and cost-effectiveness of intensive care units: A multicenter inception cohort study. *Critical Care Medicine, 22*(9), 1385–1391.

Remmlinger, E., Ault, S., & Hanrahan, L. (1995). Information technology implications of case management. *Healthcare Information Management, 9*(1), 21–28.

Report to Participants (1994). *Next-Generation Clinical Information Systems: Strategies for the Nursing Component.* Washington, DC: American Nurses Association.

Safran, C. (1993). Defining the clinical workstation. *MD Computing, 10,* 145–146.

Saul, L. (1995). Developing critical pathways: A practical guide. *Heartbeat, 5*(3), 1–13.

Schlehofer, G. B. (1992). Informatics: Managing clinical operations data. *Nursing Management, 23,* 36–38.

Shikiar, M. S., & Warner, P. (1994). Selecting financial indices to measure critical path outcomes. *Nursing Management, 25*(9), 58–60.

Shortliffe, E. H. (1994). Dehumanization of patient care - Are computers the problem or the solution? *Journal of the American Medical Informatics Association, 1,* 76–78.

Siano, B. A., & Hyland, W. P. (1994). Benchmarking in hospitals: Adapting an industry concept and achieving results. *HIMSS Proceedings, 1,* 137–152.

Tackbary, M. T. (1995). Automating critical pathways: The race is on. *HIMSS Proceedings, 4,* 379–387.

Windle, P. E. (1994). Critical pathways: An integrated documentation tool. *Nursing Management, 25*(9), 80F–80P.

Zander, K., & McGill, R. (1994). Critical and anticipated recovery paths: Only the beginning. *Nursing Management, 25*(8), 34–51.

Zielstorff, R. D., Hudgings, C. I., & Grobe, S. J. (1993). *Next-generation nursing information system: Essential characteristics for professional practice.* Kansas City, MO: American Nurses Association.

Index

Acute care critical pathways, 8, 52–53
  administrative commitment to, 53
  areas identification for, 53–54
  barriers removal and, 57
  benchmark identification and, 54–56
  conclusions regarding, 75–76
  implementation of, 73–74
  key players selection and, 54
  optimal practice definition and, 56–57
  outcomes evaluation and, 74–75
  tools and forms for use in, 57–58
    benchmarking tools, 61, 65, 68–70
    follow-up letter, 65, 72
    monitor sheet, 65, 71
    patient critical pathway, 58–59, 61,
      62–65
    standard order sheet, 58, 60–61
    structured path, 58, 59
    teaching materials, 61, 66–67
    weight chart, 65, 71–73
Administrative and operations manage-
    ment, 34
Agency for Health Care Policy and
    Research, 147
Algorithms
  computerization and, 147
  critical pathways and, 27–29, 58
  data collection for, 46
  decision-making process in, 30
  examples of, 29–31
  liability issues and, 133
Ambulatory care critical pathways, 8
  collaboration atmosphere creation and,
    80–81
  cost concentration and, 81–82
  for electroconvulsive therapy, 83, 84–85
  future trends in, 88
  health care system changes and, 78–79
  outcomes definition and, 81

for outpatient surgery, 83, 86–87
  writing the pathway and, 82–83
  See also Psychiatric home care critical
    pathways
Anxiety critical pathways, 45, 94, 101, 106
Assessment, in psychiatric home care path-
    ways, 99–101, 104
Automation. *See* Computerization

Beck Depression Inventory, 105
Benchmarking
  in acute care critical pathways, 54–56,
    61, 65, 68–70
  care process analysis and, 39
  from data collection, 19–20, 34
Breach of duty element, of malpractice lia-
    bility, 125

CABG (Coronary Artery Bypass Graph)
    critical pathway, algorithm devel-
    opment in, 29–30
Care maps, 8; *See also* Critical pathways;
    Critical pathways: design and
    implementation of
Care process analysis
  outcome data and benchmarking and, 39
  regulatory agencies and, 38
  resource utilization and, 39–40
Care provider
  outcomes on, 36–37
  variances on, 21, 42
Case management (CM)
  in acute care setting, 56
  critical pathways development and, 3, 5
  documentation management and, 35
  in psychiatric home care, 91–92
  throughout health care continuum, 140
Causation element, of malpractice liability,
    125

SP *Springer Publishing Company*

Measurement Tools in Patient Education

Barbara K. Redman, PhD, RN

This book draws together instruments for measuring outcomes in patient education from a wide variety of sources. Fifty-two actual tools are included. Each tool is accompanied by a descriptive review, a critique, and information on administration, scoring, and psychometric properties.

Measurement Tools in Patient Education

Barbara K. Redman

Springer Publishing Company

Particular attention is given to the different cultures in which the instrument has been used. Two introductory chapters give a basic orientation to the field. Designed for nurses and other health professionals involved in patient education.

Contents:

Part I: An Introduction
  • Measurement in Patient Education Practice and Research
  • Measurement of Self-Efficacy and Quality of Life in Patient Education

Part II. Description of Tools
  • Basic Patient Education Needs
  • Diabetes
  • Arthritis
  • Asthma
  • Pregnancy, Childbirth, and Parenting
  • Other Clinical Topics
  • Health Promotion, Disease Prevention and Increasing Quality of Life
  • Appendix: Summary of Tools

1997   250pp (est)   0-8261-9860-0   softcover

536 Broadway, New York, NY 10012-3955 • (212) 431-4370 • Fax (212) 941-7842

Discharge Planning For The Elderly

An Educational Program For Nurses

Kimberly Dash, MPH, **Nancy C. Zarle,** RN, MS, **Lydia O'Donnell,** EdD, and **Cheryl Vince-Whitman,** EdM

Designed for use as a text or a self-learning manual, this book offers nurses practical information on discharge planning for the elderly. The program described in the book was developed by the Education Development Center and Boston's Beth Israel Hospital. The premise of the book is that nurses should be involved in the discharge planning process, along with other members of the health care team, from the first day of hospitalization.

> Discharge Planning for the Elderly
>
> An Educational Program For Nurses
>
> Kimberly Dash
> Nancy C. Zarle
> Lydia O'Donnell
> Cheryl Vince-Whitman
>
>
>
> SPRINGER PUBLISHING COMPANY

Each chapter in this easy-to-use book includes learning objectives, case studies, and clinical exercises. The many sample forms, charts, and diagrams help to clarify complex information. Also included are worksheets for improving the discharge planning process at one's own institution.

Contents:
- Why is Discharge Planning Critical Today?
- Developing Communication Skills: The Nurse as Data Gather
- Patient Assessment
- Caregiver Assessment
- Community, Home, and Nursing Home Assessment
- Matching Clients with Available Resources
- Monitoring and Evaluating Discharge Plans
- Defining the Nurse's Role in Discharge Planning
- Case Study Development

1996   280pp   0-8261-9230-0   hardcover

536 Broadway, New York, NY 10012-3955 • (212) 431-4370 • Fax (212) 941-7842

Ⓢ *Springer Publishing Company*

Nursing Care of Geriatric Emergencies

Christine W. Bradway, RN, CS, MSN

This comprehensive guide to nursing management of medical emergencies among the elderly is for nurses who practice in acute care, long-term care, home care, and other settings. Emergencies include heart attack and heart failure, infectious disease, seizures, trauma, choking, and adverse drug reactions. Two important issues of relevance in geriatric emergencies — delirium and elder abuse and neglect — are described in separate chapters. The main emphasis is on nursing assessment and appropriate care. Discussion of epidemiology, pathophysiology, as well as case examples are included in each chapter. Care of older adults may differ significantly from what nurses traditionally learn about the emergency care of younger adults, and this book offers a comprehensive resource on these differences.

Contents:

1996   288pp   0-8261-9010-3   hardcover

536 Broadway, New York, NY 10012-3955 • (212) 431-4370 • Fax (212) 941-7842

Abandonment of the Patient
The Impact of Profit-Driven Health Care on the Public

Ellen D. Baer, RN, PhD, FAAN, **Claire M. Fagin**, RN, PhD, FAAN, and **Suzanne Gordon**

Abandonment of the Patient

The Impact of Profit-Driven Health Care on the Public

Ellen D. Baer
Claire M. Fagin
Suzanne Gordon
Editors

SPRINGER PUBLISHING COMPANY

A thoughtful and startling examination of how health care "downsizing" and for-profit managed care is affecting quality of care, patient outcomes, and staffing patterns. A distinguished group of nurses, doctors, health care administrators, patients, journalists, and policy makers have contributed to the book; among them Bruce Vladek, Arthur Caplan, and Sidney Blumenthal.

Partial Contents

- The Corporatization of American Health Care and Why It Is Happening, *Bruce C. Vladeck*
- Rescuing Quality Care: Perspective and Strategies. The Case for Managed Care, *Clark E. Kerr*
- The Case Against Profit-Driven Managed Care, *Quentin Young*
- A Historical Perspective, *Joan Lynaugh*
- Legal Nightmares/Remedies, *Ann Torregrossa*
- The Political Aftermath of Health Care Reform, *S. Blumenthal*
- Do Ethics and Money Mix? The Moral Implications of the Corporatization of Health Care, *Arthur Caplan*

1996   132pp   0-8261-9470-2   hardcover

536 Broadway, New York, NY 10012-3955 • (212) 431-4370 • Fax (212) 941-7842

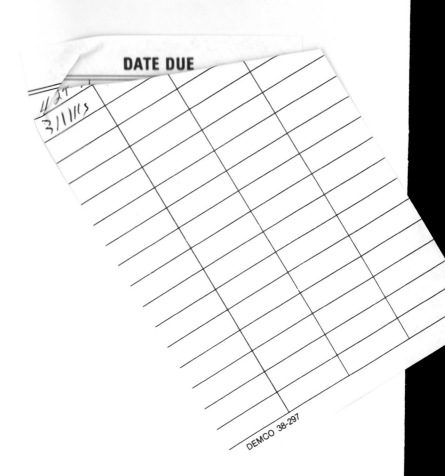

DATE DUE

DEMCO 38-297